---

APPLIED ELECTRICITY AND ELECTRONICS DIVISION

VOLUME 4

*General Editor*

P. HAMMOND

---

# LINEAR NETWORK THEORY

# LINEAR NETWORK THEORY

K. F. SANDER

UNIVERSITY LECTURER IN ENGINEERING
FELLOW OF TRINITY COLLEGE, CAMBRIDGE

PERGAMON PRESS
OXFORD · LONDON · EDINBURGH · PARIS · FRANKFURT
THE MACMILLAN COMPANY
NEW YORK

PERGAMON PRESS LTD.
Headington Hill Hall, Oxford
4 & 5 Fitzroy Square, London W.1.

PERGAMON PRESS (SCOTLAND) LTD.
2 & 3 Teviot Place, Edinburgh 1

THE MACMILLAN COMPANY
60 Fifth Avenue, New York 11, N.Y.

COLLIER-MACMILLAN CANADA, LTD.
132 Water Street South, Galt, Ontario, Canada

GAUTHIER-VILLARS ED.
55 Quai des Grands-Augustins, Paris 6

PERGAMON PRESS G.m.b.H.
Kaiserstrasse 75, Frankfurt am Main



Library of Congress Card Number: 63–23206

*Set in 10 on 12pt. Times*
*Printed in Great Britain by Blackie & Son, Ltd., Glasgow*

# CONTENTS

This book is dedicated to the memory
of J.G.Y.

# FOREWORD

NETWORK theory is inevitably an algebraic subject, and a student listening to lectures on network theory is easily reduced to taking notes of algebraic relations without appreciating the physical significance behind the façade. It has been the opinion of the author that this state of affairs could be alleviated with the aid of a book covering the algebra only, thus making it possible to give more emphasis to practically significant aspects of the theory. It is hoped that this book will provide this aid, covering the theory of network analysis as customarily taught in the last years of an honours course. No reference is made to practical circuits, although illustrations have been chosen with such circuits in mind.

The presentation commences with networks containing resistances only, and proceeds to consider networks involving inductance and capacity by way of the differential equations. Only the sinusoidal steady-state is, however, dealt with, the treatment presuming a prior acquaintance with complex methods of treating alternating currents. Formal matrix algebra is not used until the last chapter, which is a recapitulation in matrix terminology of earlier results. The term matrix is, however, used in the remainder of the text as a name for a square array of coefficients.

It gives me great pleasure to thank my colleagues for their helpful advice. I am particularly indebted to Mr. G. R. Nicoll for his detailed comments, and to Dr. D. E. Radley for working through the examples. I also wish to record the beneficial influence of many dicussions with Dr. P. R. Bryant, carried on over many years.

K. F. SANDER

# SYMBOLS

*General*

| | | |
|---|---|---|
| $R$ | resistance | used with suffices |
| $G$ | conductance | |
| $Z$ | impedance | |
| $Y$ | admittance | |

N.B. $G = 1/R$, $Y = 1/Z$ are used where convenient, without comment

| | |
|---|---|
| $M$ | mutual inductance |
| $L$ | inductance |
| $C$ | capacity |

*Chapter One*

| | |
|---|---|
| $b$ | number of branches |
| $n$ | number of nodes |
| $l$ | number of loops |
| $i_r$ | branch current |
| $u_r$ | branch voltage |
| $e_r$ | branch internal e.m.f. |
| $j_r$ | branch internal current source |
| $A_{rs}$ | elements of node-branch incidence matrix |
| $B_{rs}$ | elements of loop-branch incidence matrix |
| $v_r$ | nodal voltage |
| $i_r^{(l)}$ | loop current |
| $P_{rs}$ | elements of nodal admittance matrix |

*Chapter One*

$j_r^{(n)}$ nodal currents

$\Delta$ determinant of $P_{rs}$

$\Delta_{rs}$ cofactor of $\Delta$

*Chapter Two*

In this and all later chapters instantaneous sinusoidal quantities and the corresponding complex numbers are denoted by lower and upper case letters respectively

$x(t)$ instantaneous value of a sinusoidally varying quantity

$X$ complex representation of $x(t)$

$\left.\begin{matrix} U_r \\ I_r \\ V_r \\ J_r \end{matrix}\right\}$ complex representations of $\left\{\begin{matrix} u_r \\ i_r \\ v_r \\ j_r \end{matrix}\right.$

$\omega$ Angular frequency (often abbreviated to frequency)

$p$ $j\omega$

$*$ conjugate complex

*Chapter Three*

$z_{rs}$ elements of 2-port impedance matrix

$y_{rs}$ elements of 2-port admittance matrix

$A, B, C, D$ elements of 2-port cascade matrix

$\left.\begin{matrix} h_{rs} \\ g_{rs} \end{matrix}\right\}$ elements of 2-port hybrid matrices

*Chapter Three*

| | |
|---|---|
| $g_a, g_m, \mu$ | anode slope conductance, mutual conductance and amplification factor of a triode valve |
| $K_{rs}$ | elements of terminal admittance matrix |
| $L_{rs}$ | elements of terminal impedance matrix |

*Chapter Four*

$Z_1, Z_2, Z_3$; $P_1, P_2, P_3$; $\zeta_1, \zeta_2$ — used to denote components of T-circuit; Π-circuit; balanced lattice (respectively)

*Chapter Five*

| | |
|---|---|
| $\Gamma_1, \Gamma_2$ | iterative propagation constants |
| $\lambda, \mu$ | arbitrary constants |
| $Z_{t1}, Z_{t2}$ | iterative impedances |
| $E$ | a particular value of $\sqrt{[(A-D)^2+4BC]}$ |
| $Z_{i1}, Z_{i2}$ | image impedances |
| $T_1, T_2$ or $T_{1r}, T_{2r}$ | image propagation constants |
| $T$ | $T_1$ for a reciprocal 2-port |
| $Z_{o1}, Z_{o2}$ | open-circuit impedances for a 2-port |
| $Z_{s1}, Z_{s2}$ | short-circuit impedances for a 2-port |
| $Z_o/Z_s$ | $=Z_{o1}/Z_{s1}=Z_{o2}/Z_{s2}$ |

*Chapter Six*

| | |
|---|---|
| $\sigma^2 =$ | $\lvert Z_o/Z_s \rvert$ |
| $\alpha+j\beta$ | $=T$ |
| $K, k_1, k_2$ | reflection coefficients |

*Chapter Seven*

$$\Omega = \frac{\omega}{\omega_0} - \frac{\omega_0}{\omega}$$

*Chapter Eight*

| | |
|---|---|
| $\alpha_i, \beta_i$ | coefficients in a polynomial expansion |
| $D$ | $= \mathrm{d}/\mathrm{d}t$ |
| $\mu_i$ | roots of $\Delta(\mu) = 0$ |
| $\lambda_{rs}, a_{rs}$ | arbitrary constants |
| $F(p)$ | network transfer function |
| $H(p)$ | network function |
| $\varepsilon$ | parameter of uniform dissipation |

*Chapter Nine*

| | |
|---|---|
| $\mathscr{E}_m, \mathscr{E}_{mr}$ | instantaneous magnetic stored energy |
| $\mathscr{E}_e, \mathscr{E}_{er}$ | instantaneous electrostatic stored energy |
| $P\ (P_r)$ | instantaneous power dissipation |
| $\bar{P}$ (etc.) | time average of $P$ (etc.) |

*Chapter Ten*

| | |
|---|---|
| $\mathbf{A}_a$ | node-branch incidence matrix |
| $\mathbf{A}$ | reduced node-branch incidence matrix |
| $\mathbf{B}$ | loop-branch incidence matrix |

$$\left.\begin{matrix}\mathbf{i}\\ \mathbf{u}\\ \mathbf{v}\\ \mathbf{j}\\ \mathbf{i}^{(l)}\\ \mathbf{e}\end{matrix}\right\}\ \text{vectors of elements}\ \left\{\begin{matrix}i_r\\ u_r\\ v_r\\ j_r\\ i_r^{(l)}\\ e_r\end{matrix}\right.$$

*Chapter Ten*

- **Y** diagonal matrix of elements $G_r$
- **Z** diagonal matrix of elements $R_r$
- **P** nodal admittance matrix
- **Q** loop impedance matrix
- $\mathbf{A}^T$ etc. transpose of **A**

# 1 THE ANALYSIS OF THE GENERAL RESISTANCE-ONLY NETWORK

WHEN a number of electrical components are connected together we are said to have an electrical network. The properties of the network depend on two quite different sets of conditions: the conditions imposed by the nature of the electrical components used, and those imposed by the actual connections. The latter may be thought of as constraints on the behaviour of the components.

The complete analysis of a specific electrical network results in knowledge of the current through and voltage across each component. We may often be content with implicit knowledge of many of these quantities, requiring explicit knowledge of only one or two, and we may be said to have analysed a network when we have explicit knowledge of sufficient variables with which the remaining ones may be simply determined.

It will be found that although the answer depends, obviously, on the actual components, the method of analysis depends on the connections only. It is therefore profitable to consider networks containing sources and resistances only, before adding the other types of electrical component. Accordingly, in the present chapter, we shall deal with one electrical component and two types of source: resistances, and voltage and current sources. A resistance is defined by the property that a linear relation exists between the voltage across it and the current through it. A voltage source is defined by the property that the voltage across it is independent of the current through it, and a current source by the dual property that the current through it is independent of the voltage across it. It is at times convenient to consider the result of "removing" the sources in a network, in the sense that each voltage source is made to have zero volts across it, and each current source zero current through it. When this is done, the result is equivalent to replacing each voltage source by zero resistance, and each current source by an open circuit.

## THE GRAPH OF A NETWORK

The junctions between two or more components in a network are called nodes of the network. Each path joining a pair of nodes and through which current may flow is called a branch of the network. The pattern of interconnections is best described by a diagram, called the graph of the network, which eliminates all

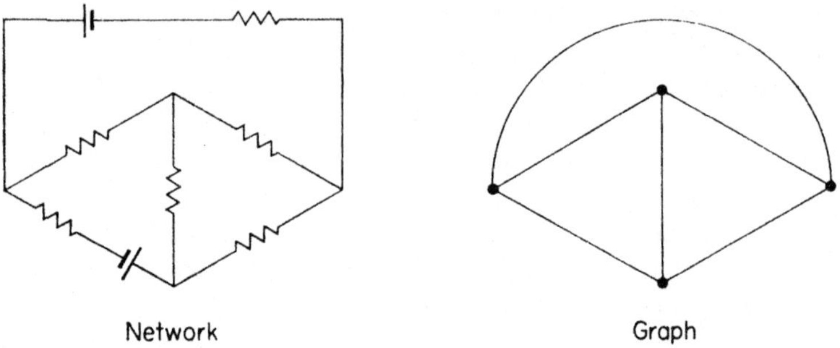


FIG. 1. *Network and graph*

the electrical properties of the components. This graph is obtained by redrawing the circuit of the network with lines replacing the electrical components. An example of a network and its graph is shown in Fig. 1.

The graph makes clear the existence of a number of closed paths which may be traced along the branches. Such closed paths are called loops. Of the total number of loops of a network, a certain number of independent loops may be chosen. One way of choosing a set of independent loops is as follows: form from the network a sub-network by removing branches until no loops remain, although each node is still connected by a single path to another node. Such a structure is called a tree of the network. Two examples for the graph of Fig. 1 are shown in Fig. 2. It

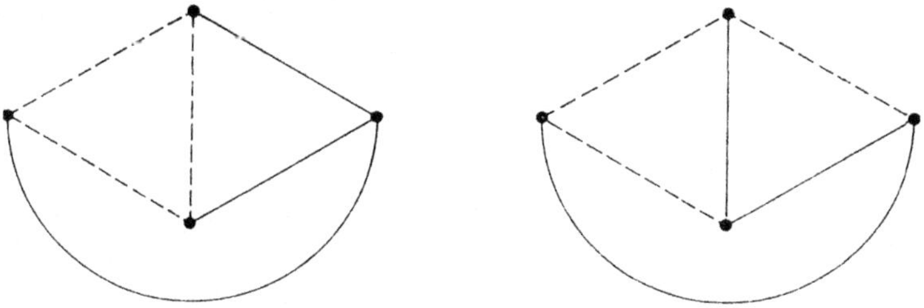

FIG. 2. *Two trees of a network*

may be seen, by starting at one node and systematically placing one branch between each pair of nodes, that if the number of nodes is $n$ the number of branches in a tree is $n-1$. Hence if the total number of branches is $b$, $b-n+1$ branches have been removed in forming the tree. These $b-n+1$ branches will be called links of the tree. Replacement of any one of these links will create a loop, and all the loops thus created are independent, since only one loop will contain a particular link. There are therefore a total number $l$ of independent loops, given by the equation

$$l = b-n+1 \tag{1}$$

This result is true however the independent loops are chosen. In the network of Fig. 1 there are three independent loops.

Although in planar networks like that of Fig. 1 it is easy to see the number of independent loops, it is not so easy when the network involves links which prevent it being opened up on to a flat surface. In such a case equation (1) provides a check, since both the nodes and the branches are easily enumerated.

## THE EQUATIONS OF THE CONSTRAINTS

In setting up the equations from which to solve the network problem, we employ the two "laws" associated with the name of Kirchhoff. These are usually stated as follows:

1. the total rise of voltage round any loop is zero;
2. the total current entering each node is zero.

The first of these results from the existence of a unique potential difference between any one node and a reference, and the second because charge cannot accumulate at a node.

Before using these laws to formulate the equations, it is necessary to make a definition of branch voltage and branch current. An actual branch is shown in Fig. 3.

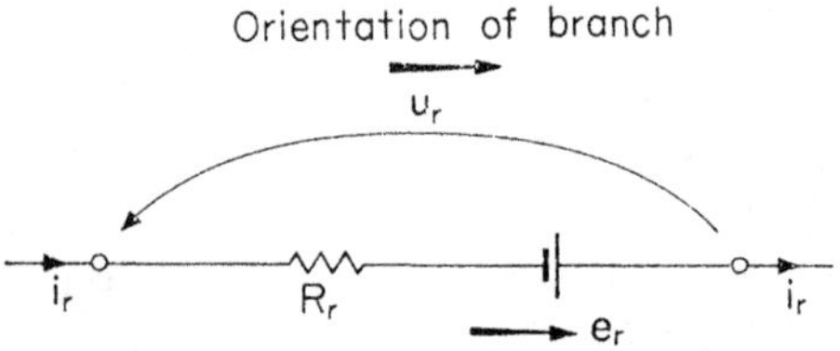


FIG. 3. *Quantities associated with the rth branch*

The branch current, $i_r$, is that current flowing into one branch terminal and out at the other. The branch voltage, $u_r$, is the voltage between branch terminals. An internal e.m.f. $e_r$ is shown. It is necessary to label the terminals of a branch when defining the sign of these quantities. Once the positive sense of $u_r$ has been defined, the direction of $i_r$ is as shown in Fig. 3. The addition of terminals to the branch and the definition of branch current outside those terminals avoids ambiguity when a current generator is present, as shown in Fig. 4.

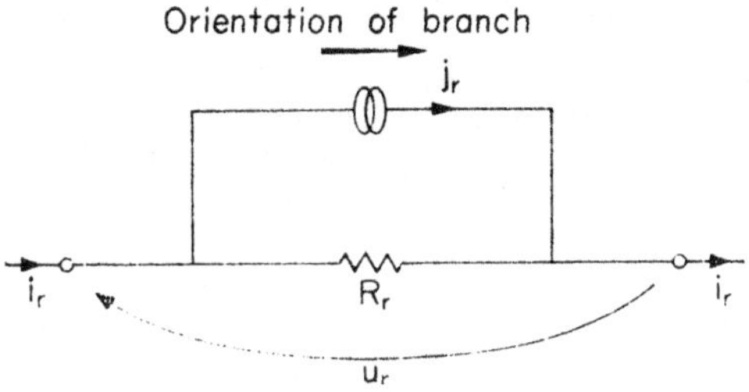


FIG. 4. *Branch with current generator*

In general, both types of generator may be present, but consideration of each separately gives a simpler treatment. The establishment in a particular network of numbering for the branches and of convention for the sense of branch voltages enables the laws to be used to formulate two sets of algebraic equations. From the first law we have, choosing independent loops, a set of equations $l$ in number, of the form

$$\sum_{s=1}^{b} B_{rs} u_s = 0, \qquad r = 1, 2, \ldots, l \tag{2}$$

The coefficients $B_{rs}$ are all either $\pm 1$ or 0, depending whether the $s$th branch is in the loop being considered, for some specified $r$, and on the relation between the sense of $u_s$ and the direction round the loop. The orientation of a branch will be defined to be that of the current through it, and $B_{rs} = +1$ if the direction of the $s$th branch coincides with that of the loop.

The second law yields $n$ equations of the form

$$\sum_{s=1}^{b} A_{rs} i_s = 0, \qquad r = 1, 2, \ldots, n \tag{3}$$

The coefficients $A_{rs}$ are zero if branch $s$ does not begin or end on node $r$, $+1$ if branch $s$ begins on node $r$, and $-1$ if branch $s$ ends on node $r$. It has been assumed that all the sources are internal to the network, and hence if current is conserved at $n-1$ nodes it is automatically conserved at the remaining one. Hence only $n-1$ of equations (3) are independent. Taking equations (2) and (3) together we have a total of $l+n-1 = b$ equations relating the $2b$ unknowns $u_r$ and $i_r$, $(r = 1,2,\ldots,b)$, leaving a further $b$ relations to be specified. These are obtained by applying Ohm's law to the branches. For the branches of Figs. 3 and 4 we have respectively

$$\left.\begin{aligned} u_r &= R_r i_r - e_r \\ \text{or} \quad u_r &= R_r(i_r - j_r) \end{aligned}\right\} \quad r = 1,2,\ldots,b \tag{4}$$

We thus have a total of $2b$ equations in $2b$ unknowns.

## REDUCTION OF NUMBER OF UNKNOWNS

Equations (2), (3) and (4) are sufficient to determine all branch voltages and currents, but it is not usually necessary to find them all. A lesser number of unknowns may be obtained by manipulation of the equations and change of variable. There are two generally used ways, although others certainly exist. These two arise by automatically satisfying either equations (2) or (3) by choice of unknowns.

## NODAL VOLTAGES

Equations (2) may be satisfied automatically if the branch voltages are expressed in terms of $n-1$ nodal voltages. One node, say node $n$, is taken as origin and $n-1$ unknowns $v_r$ defined so that $v_r$ is the voltage of node $r$ with respect to node $n$. Each branch voltage will be the difference between two nodal voltages. If expressions for branch voltages in terms of $v_r$ are substituted in equations (4), expressions for the branch currents in terms of $v_r$ will be obtained. These in turn may be substituted in equations (3), yielding finally $n-1$ equations in $n-1$ unknowns.

## LOOP CURRENTS

An alternative procedure is to associate with each of the $l$ independent loops chosen, loop currents $i_1^{(l)}, i_2^{(l)}, \ldots, i_l^{(l)}$ thought of as flowing round each loop in a prescribed direction. Each branch is contained in one or more loops and the branch currents may be expressed as the sum or difference of a number of loop currents. The loop currents evidently satisfy conservation at each node. Using equations (4) to obtain expressions for the branch voltages and then substituting these expressions into equations (2) yields a set of $l$ equations in the $l$ unknowns $i_1^{(l)}, i_2^{(l)}, \ldots, i_l^{(l)}$.

## COMMENTS ON NODAL AND LOOP METHODS

1. It is evident that the numbers of unknowns involved are different in the two cases, and that therefore for any given problem one is preferable to the other, depending on the relative magnitudes of $n-1$ and $l$.

It may be observed that in many cases one might expect the nodal method to involve fewer unknowns. This is because most practical networks have many cross connections and resemble the "fully connected" network. This is defined as a network in which each node is connected by a component to all the others. The number of branches in such a network of $n$ nodes is

$$(n-1)+(n-2)+(n-3)+\ldots+1 = \tfrac{1}{2}n(n-1)$$

The number of independent loops is therefore

$$\tfrac{1}{2}n(n-1)-n+1 = (n-1)(\tfrac{1}{2}n-1)$$

The number of independent loops is greater than $n-1$ if $n \geqq 5$. For a moderately well-connected network it may easily happen that the number of loops exceeds $n-1$, and in this case the nodal method would involve fewer unknowns.

2. The nodal method is often described as being necessarily associated with current generators, so that before applying the method it is thought necessary to write the network in terms of current generators only.

This is quite unnecessary and for simple circuits worked from first principles the voltage sources may be left in. It is true of course, as will be seen later, that for the purposes of systematic analysis, it is more convenient to use current generators in nodal methods and voltage generators with loop methods.

3. In a complicated circuit nodes are much more easily visualised than loops. This sometimes offers a reason for working with nodal analysis.

4. For formal solutions it is quite irrelevant which system is used. Each system must yield the same results. In this book nodal methods will be used.

## THE FORMAL NODAL EQUATIONS

Equations (2), (3) and (4) may be used to arrive at the formal equations in terms of nodal variables or loop variables. This process is most easily, but not necessarily, done by matrix methods. These will not be used in the present chapter, but they are outlined in Chapter 10. We shall take nodal voltages for unknowns and set up the equations directly.

Consider that part of a circuit shown in Fig. 5, in which all connections to node 2 are shown.

The equation of conservation of current at node 2 is

$$i_1+i_2-i_3 = 0 \tag{5}$$

In terms of nodal voltages $v_1$, $v_2$ and $v_3$ we have

$$i_1 = \frac{1}{R_1}(-v_2+v_1+e_1)$$

$$i_2 = \frac{1}{R_2}(-v_2+e_2)$$

$$i_3 = \frac{1}{R_3}(-v_3+v_2-e_3)$$

It is obviously more satisfactory to deal in conductances than in resistances, and if the conductance $1/R_r$ be denoted by $G_r$, equation (5) becomes

$$-G_1 v_1+(G_1+G_2+G_3)v_2-G_3 v_3 = G_1 e_1+G_2 e_2+G_3 e_3 \qquad (6)$$

The significance of the individual coefficients is simple: the first and third are the negative of the conductance of the resistor joining node 1 to node 2 and node 3 to node 2 respectively. The second coefficient is the total conductance between node 2 and

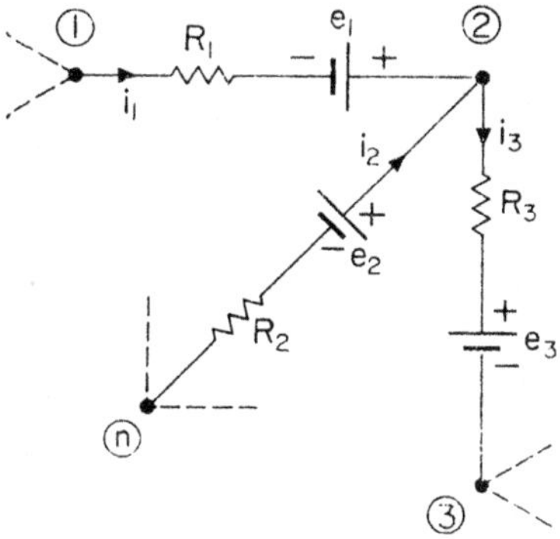


FIG. 5. *Part of a network*

all other nodes. The right-hand side represents the total current flowing into node 2 from the branches if all nodes are joined together.

This last term may be given a perhaps more physical significance if the network is redrawn as in Fig. 6, using equivalent current generators.

The total current flowing into the node *from the generators* is seen to be the right-hand side of equation (6). It must be noted that the transformation between Figs. 5 and 6 will give the same branch currents and voltages in the sense defined earlier. The currents in the resistors will, however, be different in the two circuits. The nodal voltages found for the circuit of Fig. 6 may

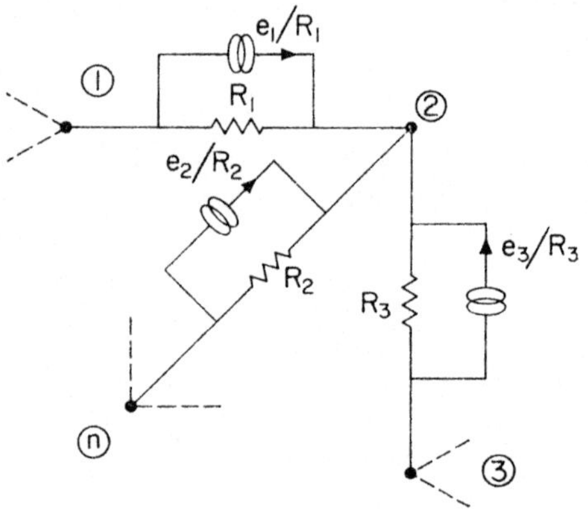


FIG. 6. *Network with equivalent current generators*

be used in that of Fig. 5 to find, for example, the power dissipated in an individual resistor, but this power will not necessarily be the same as that dissipated in the same resistor in the circuit of Fig. 6.

It is straightforward to set up equations similar to equation (6) for all the other nodes. The general set may be written in the form

$$\left.\begin{aligned} P_{11}v_1+P_{12}v_2+\ldots+P_{1,n-1}v_{n-1} &= j_1^{(n)} \\ P_{21}v_1+P_{22}v_2+\ldots+P_{2,n-1}v_{n-1} &= j_2^{(n)} \\ P_{n-1,1}v_1+P_{n-1,2}v_2+\ldots+P_{n-1,n-1}v_{n-1} &= j_{n-1}^{(n)} \end{aligned}\right\} \quad (7)$$

Each term may be given a significance parallel to that in equation (6). The coefficients $P_{kl}$, $k \neq l$, are equal to the negative of the conductance joining nodes $k$ and $l$. It follows automatically that $P_{kl}=P_{lk}$. The coefficient $P_{kk}$ is the total conductance joining node $k$ to all the other nodes joined together. It follows that

$$\sum_{l=1}^{n-1} P_{kl} \geqq 0$$

The term $j_k^{(n)}$ is equal to the total current flowing into node $k$ from the branches when all the other nodes are joined together.

## SOLUTION OF THE EQUATIONS

Two types of solution are relevant depending on whether the problem is to provide a numerical answer or to provide a general theoretical statement.

If numerical data is provided the equations are best solved by a method of systematic elimination, using, for example, the first equation to eliminate $v_1$ from all the rest. The first of the resulting equations is used to eliminate $v_2$ and so on, finally leaving one equation in the last unknown. It is desirable to order the equations in the best manner, but it seems unnecessary to dwell on this point here, since most cases where such refinement is necessary will be dealt with by expert computers.

If theoretical results are required, recourse must be made to the theory of determinants. The formal solution of equations (7) is given by

$$v_l = \sum_{k=1}^{n-1} \frac{\Delta_{lk}}{\Delta} j_k^{(n)} \tag{8}$$

in which $\Delta$ is the determinant

$$\begin{vmatrix} P_{11} & P_{12} & \cdot & \cdot & P_{1,n-1} \\ P_{21} & P_{22} & \cdot & \cdot & P_{2,n-1} \\ \cdot & \cdot & \cdot & \cdot & \cdot \\ P_{n-1,1} & P_{n-1,2} & \cdot & \cdot & P_{n-1,n-1} \end{vmatrix}$$

and $\Delta_{lk}$ is the determinant obtained from $\Delta$ by deleting the $l$th row and $k$th column. Providing the reference node is connected to the network $\Delta$ will be non-zero.

The general properties of network determinants of this type are very necessary in the theory of network synthesis. It will however be adequate for the purpose of this book to observe that the determinant $\Delta$ may be expanded in the form

$$\Delta = \sum (\pm) P_{1\alpha} P_{2\beta} \dots P_{n-1,\nu} \tag{9}$$

the sum being taken over all possible combinations of different integers between 1 and $n-1$ for the $n-1$ indices $\alpha, \beta, \dots, \nu$. The sign for each term is determined by the oddness or evenness of the permutation of the integers, but for the present purpose it is necessary to consider only the nature of the terms and not their numerical value.

That equations (8) must take the form dictated by this expansion and the parallel one for $\Delta_{lk}$ is obvious from consideration of any process of elimination of variables from the simultaneous equations (7).

## RECIPROCITY

For the type of network considered it was pointed out that $P_{kl} = P_{lk}$. This implies from the form of $\Delta$ that $\Delta_{lk} = \Delta_{kl}$.

An alternative formulation of this result may be obtained directly from the equations (7). Consider the same network under two different sets of driving currents $j_k^{(n)} = J'_k, J''_k$ $(k = 1, 2, \dots, n-1)$ and let the corresponding nodal voltages be $v'_l, v''_l$ $(l = 1, 2, \dots, n-1)$. We have

$$\sum_{k=1}^{n-1} P_{lk} v'_k = J'_l \tag{10}$$

$$\sum_{k=1}^{n-1} P_{lk} v''_k = J''_l$$

Hence
$$\sum_{k=1}^{n-1} v'_l J''_l = \sum_{l=1}^{n-1} v'_l \sum_{k=1}^{n-1} P_{lk} v''_k$$
$$= \sum_{k=1}^{n-1} v''_k \sum_{l=1}^{n-1} P_{lk} v'_l$$

Now since
$$P_{lk} = P_{kl}$$
$$\sum_{l=1}^{n-1} P_{lk} v'_l = \sum_{l=1}^{n-1} P_{kl} v'_l$$

Using the first of equations (10) this may be seen to be equal to $J'_k$.

Hence
$$\sum_{l=1}^{n-1} v'_l J''_l = \sum_{l=1}^{n-1} v''_l J'_l \tag{11}$$

Direct substitution for $v'_l$ from equation (8) into the left-hand side of equation (11) will lead to the same identity.

We may deduce the property of resistance-only networks known as reciprocity. Consider the two configurations shown in Fig. 7(*a*) and 7(*b*), in which only nodes 1, 3, 4, and *n* are shown, and in which all other injected currents other than those shown are zero.

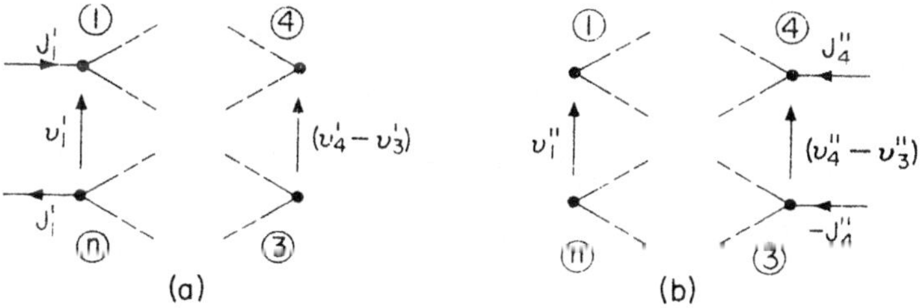


FIG. 7. *Experiments illustrating reciprocity*

In each case we are concerned with two pairs of terminals with excitation applied first to one and then to the other.

Applying equation (11) we have

$$v_3'(-)J_4''+v_4'J_4'' = v_1''J_1'$$

or

$$\frac{v_4'-v_3'}{J_1'} = \frac{v_1''}{J_4''} \tag{12}$$

This equation relates the voltage between one pair of terminals to the current flowing between another pair and shows that the ratio is independent of the end to which the voltage is applied. A network which obeys this relation is said to be *reciprocal.*

*Examples on Chapter 1*

1. Determine the number of different trees of the graph below. *Answer:* 8.

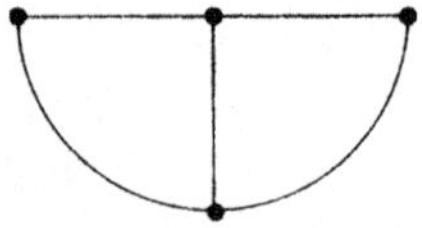

2. Determine the number of independent loops by inspection and by means of equation (1) for the networks whose graphs are
(*a*) a cube,
(*b*) a hexagon *ABCDEF* with branches joining *AD*, *BE* and *CF*.

3. For the circuit, numbering and directions of the figure, set up the array of coefficients $A_{rs}$, $B_{rs}$.

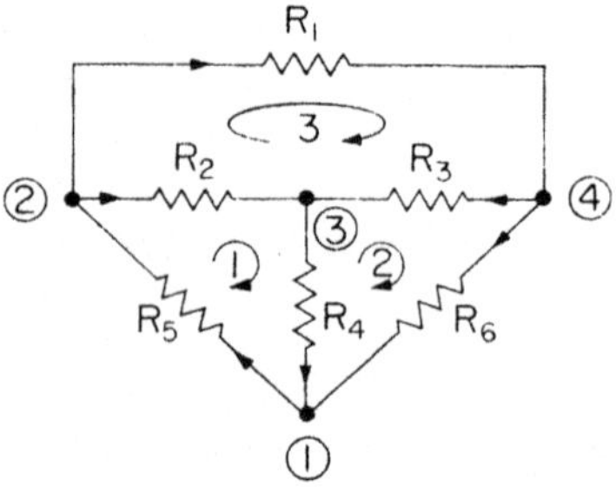

*Answer:*

$$A_{rs} = \downarrow_{r} \overset{s \rightarrow}{\begin{bmatrix} 0 & 0 & 0 & -1 & 1 & -1 \\ 1 & 1 & 0 & 0 & -1 & 0 \\ 0 & -1 & -1 & 1 & 0 & 0 \\ -1 & 0 & 1 & 0 & 0 & 0 \end{bmatrix}}$$

$$B_{rs} = \downarrow_{r} \overset{s \rightarrow}{\begin{bmatrix} 0 & 1 & 0 & 1 & 1 & 0 \\ 0 & 0 & -1 & -1 & 0 & 1 \\ 1 & -1 & 1 & 0 & 0 & 0 \end{bmatrix}}$$

4. Assuming that each branch in the circuit of question 3 contains a source of e.m.f. directed with the branch, obtain expressions for the branch currents in terms of three loop currents $I_1$, $I_2$ and $I_3$. Hence set up equations for the loop currents.

*Answer:*

$$(R_2+R_4+R_5)I_1 - R_4 I_2 - R_2 I_3 = e_2+e_4+e_5$$
$$-R_4 I_1 + (R_3+R_4+R_6)I_2 - R_3 I_3 = -e_3 - e_4 + e_6$$
$$-R_2 I_1 - R_3 I_2 + (R_1+R_2+R_3)I_3 = e_1 - e_2 + e_3$$

5. Assuming each branch in the circuit of question 3 to contain a current generator in parallel, with the conventions of Fig. 4, obtain expressions for branch currents in terms of nodal voltages $v_1$, $v_2$ and $v_3$ measured relative to node 4. Hence set up equations for these nodal potentials.

*Answer:*

$$(G_4+G_5+G_6)v_1 - G_5 v_2 - G_4 v_3 = j_4 - j_5 + j_6$$
$$-G_5 v_1 + (G_1+G_2+G_5)v_2 - G_2 v_3 = -j_1 - j_2 + j_5$$
$$-G_4 v_1 - G_2 v_2 + (G_2+G_3+G_4)v_3 = j_2 + j_3 - j_4$$

6. Determine the nodal voltage $v$ for the two "equivalent" circuits given below, and compare expressions for the power dissipated in the resistance $R_1$ in the two cases.

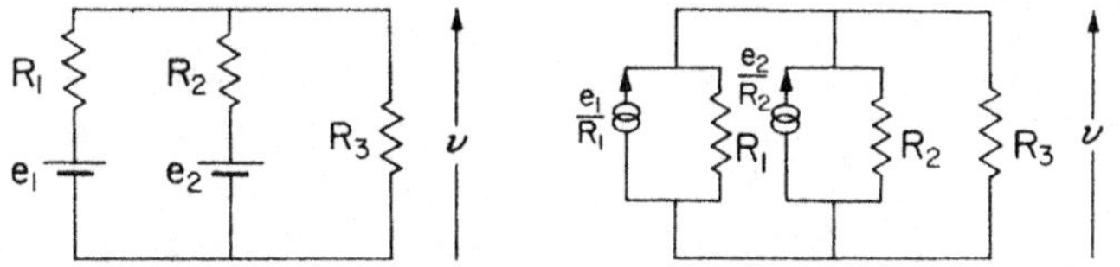

7. For the network below, write down the coefficients $P_{kl}$ of equation (7), nodal voltages being referred to node 5.

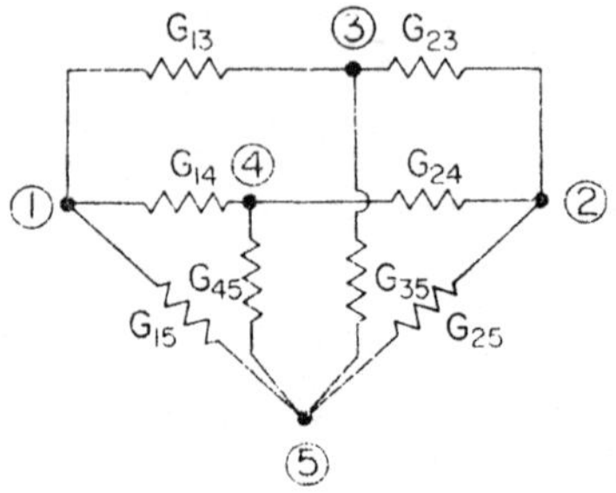


*Answer:*

$$\begin{bmatrix} G_{13}+G_{14}+G_{15} & 0 & -G_{13} & -G_{14} \\ 0 & G_{23}+G_{24}+G_{25} & -G_{23} & -G_{24} \\ -G_{13} & -G_{23} & G_{13}+G_{23}+G_{35} & 0 \\ -G_{14} & -G_{24} & 0 & G_{14}+G_{24}+G_{45} \end{bmatrix}$$

8. The coefficient $P_{kl}$ for the network of question 3 is furnished by the solution to question 5. If $G_1 = 0$, the network has the graph of question 1, in which the number of different trees was found. For each tree take the product of the conductances of the branches in the tree, and sum the products for all trees. Show this sum is equal to the determinant of the coefficients $P_{kl}$.

*Answer:* $$G_2 G_3 G_4 + G_4 G_5 G_6 + G_2 G_4 G_6 + G_2 G_3 G_6 + \\ + G_2 G_5 G_6 + G_3 G_5 G_6 + G_3 G_4 G_5 + G_2 G_3 G_5$$

(This is generally true, and is one of what are known as "Kirchhoff's rules".)

9. Consider two nodes, 1 and 2, of a network which can have currents $J_1$, $J_2$ injected. In one experiment no current is fed to or drawn from node 2, and the ratio of the potential of node 1 to that of node 2 is found. Show that if node 1 is joined to the

reference node, the ratio of the current injected at node 2 to that drawn from node 1 is given by the previous ratio.

10. Show that, if node $n$ is taken as reference for nodal voltages, the branch voltages are expressed in terms of the nodal voltages according to the equations

$$u_r = \sum_{s=1}^{n-1} A_{sr} v_s$$

Show also that

$$\sum_{r=1}^{b} u_r i_r = 0$$

11. A network is composed of branches of the type of Fig. 3. If single and double primes denote different experiments with the network prove that

$$\sum_{r=1}^{b} e'_r i''_r = \sum_{r=1}^{b} e''_r i'_r$$

# 2 *RLC* NETWORKS

THE results derived in Chapter 1 are true for any time-varying branch voltages and currents. Equations (2) and (3) of that chapter depend only on the constraints, i.e. the network connections, and are valid for any instantaneous quantities possible in the network. It is only the branch equations which require changing to deal with other types of component such as inductance and capacity.

Three branch types may be distinguished, as shown in Fig. 1. In this figure the generators are shown as current generators in parallel with the component. Any voltage generator can be so represented. To prove this consider an inductance in series with a source of e.m.f. $e(t)$. In accordance with the conventions used

$$u = L\frac{\mathrm{d}i}{\mathrm{d}t} - e$$

$$= L\frac{\mathrm{d}}{\mathrm{d}t}\left[i - \frac{1}{L}\int_{-\infty}^{t} e\,\mathrm{d}t\right]$$

This combination is therefore equivalent in terminal behaviour to the circuit of Fig. 2.

It must be emphasised that it is only in terminal behaviour that the two circuits are identical. The stored energy, for example, will not be the same.

In the case of a capacity the current generator equivalent to a series e.m.f. will be $C\mathrm{d}e/\mathrm{d}t$. If $e$ is permitted to have discontinuities, impulse functions of the equivalent current will result.

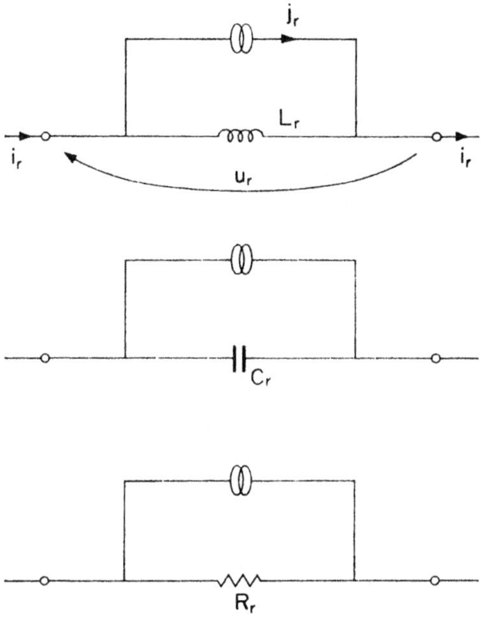


FIG. 1. *Inductive, capacitative and resistive branches*

Mathematically the final result will be the same, but care will be needed in applying initial conditions. In the present context such difficulties will not arise since only the sinusoidal steady-state will be considered.

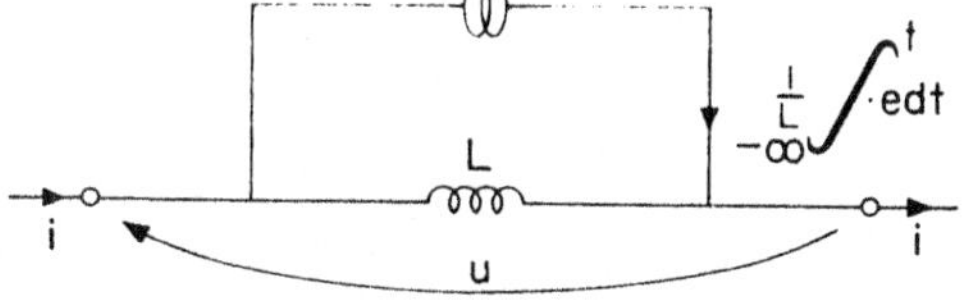


FIG. 2. *Equivalent branch*

## THE SINUSOIDAL STEADY-STATE

If all generators have a time variation which is sinusoidal and of the same frequency, the differential equations may be transformed into algebraic equations with complex coefficients. This transformation may be effected in a number of ways. We may firstly consider the identity

$$\begin{aligned} X_0 \cos(\omega t+\alpha) &= \tfrac{1}{2}(X_0 \mathrm{e}^{j\alpha}.\mathrm{e}^{j\omega t}+X_0 \mathrm{e}^{-j\alpha}.\mathrm{e}^{-j\omega t}) \\ &= \mathrm{Re}(X_0 \mathrm{e}^{j\alpha}.\mathrm{e}^{j\omega t}) \end{aligned}$$

We may therefore associate with the expression

$$x(t) = X_0 \cos(\omega t+\alpha) \tag{1}$$

a complex number $X = X_0 \mathrm{e}^{j\alpha}$ (2)

If it is given that the frequency is $\omega/2\pi$ the expressions (1) and (2) are equivalent and related by

$$x(t) = \tfrac{1}{2}(X \mathrm{e}^{j\omega t}+X^* \mathrm{e}^{-j\omega t}) = \mathrm{Re}(X \mathrm{e}^{j\omega t}) \tag{3}$$

in which the asterisk denotes the complex conjugate.

The equation describing the first branch illustrated in Fig. 1 is

$$u_r = L_r \frac{\mathrm{d}i_r}{\mathrm{d}t} - L_r \frac{\mathrm{d}j_r}{\mathrm{d}t} \tag{4}$$

If the variables $u_r$, $i_r$ and $j_r$ are expressed after the manner of equation (3) by complex numbers $U_r$, $I_r$ and $J_r$ respectively, equation (4) becomes

$$\begin{aligned} U_r \mathrm{e}^{j\omega t}+U_r^* \mathrm{e}^{-j\omega t} = L_r \frac{\mathrm{d}}{\mathrm{d}t}(I_r \mathrm{e}^{j\omega t}+I_r^* \mathrm{e}^{-j\omega t}) - \\ - L_r \frac{\mathrm{d}}{\mathrm{d}t}(J_r \mathrm{e}^{j\omega t}+J_r^* \mathrm{e}^{-j\omega t}) \end{aligned}$$

This equation is satisfied if the equation

$$U_r e^{j\omega t} = L_r \frac{d}{dt}(I_r e^{j\omega t}) - L_r \frac{d}{dt}(J_r e^{j\omega t})$$

is satisfied. This yields the algebraic equation

$$U_r = j\omega L_r (I_r - J_r) \tag{5}$$

$j\omega L_r$ is known as the impedance of $L_r$ at frequency $\omega/2\pi$.

The equations expressing the constraints imposed by the network are typified by such equations as equation (3) of Chapter 1.

$$\sum_{s=1}^{b} A_{rs} i_s = 0, \qquad r = 1, 2, \ldots, n \tag{6}$$

If $i_s$ has a sinusoidal waveform and is represented by the complex number $I_s$ these equations become

$$\sum_{s=1}^{b} A_{rs}(I_s e^{j\omega t} + I_s^* e^{-j\omega t}) = 0$$

If this is true for all time

$$\sum_{s=1}^{b} A_{rs} I_s = \sum_{s=1}^{b} A_{rs} I_s^* = 0$$

The two sums are complex conjugates since the coefficients $A_{rs}$ are real, and hence equations (6) may be replaced by

$$\sum_{s=1}^{b} A_{rs} I_s = 0 \tag{7}$$

## FOURIER TRANSFORM METHOD

A method which gives the same result but which introduces the idea of transformation rather than replacement is to use a simple case of the Fourier transform. Let $x(t)$ be a sinusoidal waveform as expressed by equation (1). A one-to-one relationship exists between $x(t)$ and the number

$$X = \frac{1}{2} \times \frac{\omega}{2\pi} \int_{-\pi/\omega}^{\pi/\omega} x(t)\, e^{-j\omega t}\, dt \tag{8}$$

The factor of $\frac{1}{2}$ has been added to make $X$ identical with that of equation (2), but is otherwise irrelevant. In fact

$$x(t) = \tfrac{1}{2}(X\, e^{j\omega t} + X^*\, e^{-j\omega t}) \tag{9}$$

This is verified by the work of the preceding sections.

We may say the operation of integrating $x(t)$ multiplied by $(\omega/4\pi)\exp(-j\omega t)$ between $\pm\pi/\omega$ transforms $x(t)$ to $X$. Further forming the sum in equation (9) transforms $X$ into $x(t)$. It is evident that any linear relation between quantities having this time variation can be transformed by this operation. The application to equation (4) yields

$$\frac{\omega}{4\pi} \int_{-\pi/\omega}^{\pi/\omega} u_r(t)\, e^{-j\omega t}\, dt = L_r \frac{\omega}{4\pi} \int_{-\pi/\omega}^{\pi/\omega} \frac{d}{dt}(i_r - j_r)\, e^{-j\omega t}\, dt$$

Integration by parts gives

$$\frac{\omega}{4\pi} \int_{-\pi/\omega}^{\pi/\omega} u_r(t)\, e^{-j\omega t}\, dt$$

$$= \frac{\omega}{4\pi} \left[ L_r \{(i_r - j_r)\, e^{-j\omega t}\}_{-\pi/\omega}^{\pi/\omega} + j\omega L_r \int_{-\pi/\omega}^{\pi/\omega} (i_r - j_r)\, e^{-j\omega t}\, dt \right]$$

The expression between limits vanishes since $i_r$ and $j_r$ are periodic in $2\pi/\omega$, and hence denoting transforms by capitals we arrive at

$$U_r = j\omega L_r (I_r - J_r)$$

which is identical with equation (5). Hence the operation giving $X$ from $x(t)$ transforms the differential equation into a linear algebraic equation. The transformation applied to equations (6) evidently yields equations (7) immediately.

Although this transform method gives the same result as what may be termed the method of algebraic substitution the idea of a transform is capable of wide generalisation, and this alternative way of looking at the standard process is an introduction to more general applications.

Similar arguments apply to the other branches shown in Fig. 1 and the transformed branch equations are

$$U_r = \frac{1}{j\omega C_r}(I_r - J_r)$$

$$U_r = R_r(I_r - J_r)$$

The coefficients give expression for the impedances of the components at the frequency $\omega/2\pi$.

## A GENERAL BRANCH

In advanced network theory it is often desirable to treat each single component as constituting a branch. Within the scope of the present book it will not be necessary to do this and it will suffice to take a general branch of the form shown in Fig. 3. In

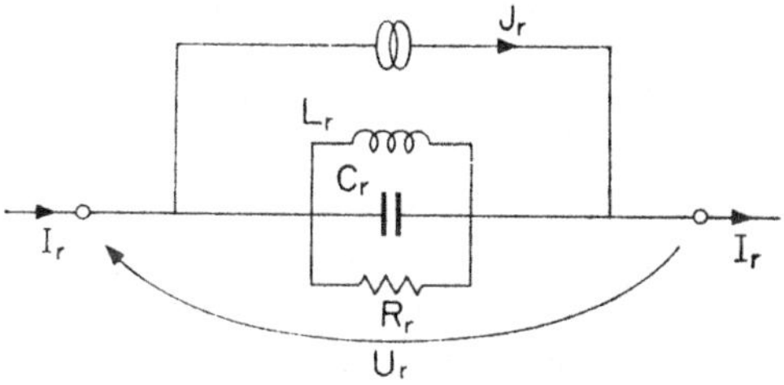


FIG. 3. *General branch for steady-state variables*

this diagram the complex notation has replaced the instantaneous values, and henceforward the steady-state sinusoidal case will be implied. The equation for this branch will be written

$$I_r = J_r + \left(\frac{1}{j\omega L_r} + j\omega C_r + \frac{1}{R_r}\right) U_r$$

Since the combination $j\omega$ is the only one in which $j$ occurs, it is algebraically convenient to write†

$$j\omega = p$$

This gives the final branch equation

$$I_r = J_r + Y_r(p)U_r \tag{10}$$

where

$$Y_r(p) = \frac{1}{L_r p} + C_r p + \frac{1}{R_r} \tag{11}$$

This expression is of course the admittance of the parallel $L$, $C$, $R$ circuit.

The general parallel branch has been chosen because it is intended to formulate the network equation in nodal form. It is equally possible to take a general series circuit with a series source of e.m.f. It is also possible to transform the circuit of Fig. 3 to the same impedance in series with an equivalent e.m.f.

## THE FORMAL NETWORK EQUATIONS

The process adopted in Chapter 1 may now be repeated. The equations are identical in mathematical form except that the branch equation has $Y_r(p)$ instead of $G_r$. Nodal voltages can be defined with respect to node $n$ and the branch voltages expressed in terms of these. The final result is to obtain a set of equations of the form of equations (7) of Chapter 1.

$$\left.\begin{aligned} P_{11}V_1 + P_{12}V_2 + \ldots + P_{1,n-1}V_{n-1} &= J_1 \\ P_{21}V_1 + P_{22}V_2 + \ldots + P_{2,n-1}V_{n-1} &= J_2 \\ \\ P_{n-1,1}V_1 + P_{n-1,2}V_2 + \ldots + P_{n-1,n-1}V_{n-1} &= J_{n-1} \end{aligned}\right\} \tag{12}$$

The coefficients have the same significance: $P_{rs}$, $r \neq s$, is the nega-

† Since $p$ has other connotations, it must be emphasised that this is solely an algebraic change of variable. In this book $p$ denotes an algebraic variable which is imaginary for real frequencies.

tive of the admittance joining node $r$ to node $s$; $P_{rr}$ is the total admittance between node $r$ and all other nodes joined together. It is to be noted once again that $P_{rs} = P_{sr}$.

The solution of the equations is identical to the resistance-only case except that the determinants involved are now functions of $p$. Hence they will be complex numbers for real frequencies. The behaviour of the determinants as functions of $p$ is of great importance, but consideration of this aspect will be deferred until Chapter 8.

The array

$$\begin{bmatrix} P_{11} & P_{12} & & \cdot & P_{1,n-1} \\ P_{21} & P_{22} & \cdot & \cdot & P_{2,n-1} \\ \cdot & \cdot & \cdot & \cdot & \cdot \\ P_{n-1,1} & P_{n-1,2} & \cdot & \cdot & P_{n-1,n-1} \end{bmatrix}$$

is called the nodal admittance matrix for the network. Although the methods of matrix algebra are confined to Chapter 10, it is convenient to have the term "matrix" for this array of coefficients.

## PARALLELING OF NETWORKS

If two networks each of $n$ nodes are paralleled by joining one node of one network to a corresponding node of another, it is obvious from the definition of $P_{rs}$ that the nodal admittance matrix of the resultant network is obtained by forming the sums

$$\begin{bmatrix} P'_{11}+P''_{11} & P'_{12}+P''_{12} & \cdot\ \cdot & P'_{1,n-1}+P''_{1,n-1} \\ \cdot & \cdot & \cdot\ \cdot & \cdot \\ P'_{n-1,1}+P''_{n-1,1} & P'_{n-1,2}+P''_{n-1,2} & \cdot\ \cdot & P'_{n-1,n-1}+P''_{n-1,n-1} \end{bmatrix}$$

in which the single and double primes refer to the two networks.

It is not necessary that both should be of the same number of nodes, for a number of unconnected nodes may be added to each network, the total number of nodes being the same. The coefficients $P_{rs}$ belonging to these unconnected nodes will be zero.

## MUTUAL INDUCTANCE

The analysis of a pair of coupled coils is well known, and in terms of the variables shown in Fig. 4

$$V_1 - V_2 = pL_1 I_1 + pMI_3$$
$$V_3 - V_4 = pMI_1 + pL_2 I_3$$

If these equations are solved for $I_1$ and $I_3$ we obtain

$$\left.\begin{aligned} I_1 &= \frac{1}{p(L_1 L_2 - M^2)}\{L_2 V_1 - L_2 V_2 - MV_3 + MV_4\} \\ I_3 &= \frac{1}{p(L_1 L_2 - M^2)}\{-MV_1 + MV_2 + L_1 V_3 - L_1 V_4\} \end{aligned}\right\} \quad (13)$$

These equations relate injected currents at nodes 1 and 3 to the nodal voltages. To complete the description we should add two further equations bringing in the currents injected at nodes 2 and 4, $-I_1$, $-I_3$ respectively. These are obtained by reversing the signs of equations (13). The final equations are therefore

$$I_1 = \frac{1}{p(L_1 L_2 - M^2)}\{L_2 V_1 - L_2 V_2 - MV_3 + MV_4\}$$

$$-I_1 = I_2 = \frac{1}{p(L_1 L_2 - M^2)}\{-L_2 V_1 + L_2 V_2 + MV_3 - MV_4\}$$

$$I_3 = \frac{1}{p(L_1 L_2 - M^2)}\{-MV_1 + MV_2 + L_1 V_3 - L_1 V_4\}$$

$$-I_3 = I_4 = \frac{1}{p(L_1 L_2 - M^2)}\{MV_1 - MV_2 - L_1 V_3 + L_1 V_4\}$$

These equations are in the form of equations (12) with a nodal admittance matrix

$$\frac{1}{p(L_1 L_2 - M^2)} \begin{bmatrix} L_2 & -L_2 & -M & M \\ -L_2 & L_2 & M & -M \\ -M & M & L_1 & -L_1 \\ M & -M & -L_1 & L_1 \end{bmatrix} \tag{14}$$

It will be noticed that this matrix is symmetrical, i.e. $P_{rs} = P_{sr}$. Hence if the network of Fig. 4 is added in parallel to any four

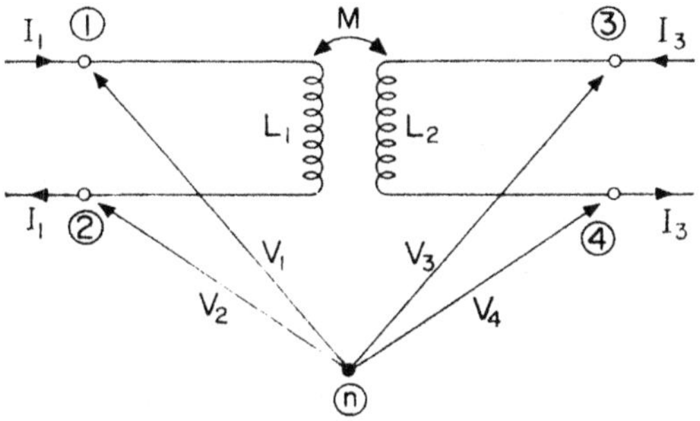


FIG. 4. *Mutual inductance*

nodes of an *RLC* network, the resulting network is still reciprocal.

It may be noticed that the sum of each row of the matrix of equation (14) is zero. This is because, in Fig. 4, the reference node is not connected to one of the nodes of the network being considered. When such a reference system is used the resulting matrix is called the indefinite nodal admittance matrix.

## IDEAL TRANSFORMERS

The arguments of the previous section break down if perfect coupling exists between $L_1$ and $L_2$, i.e. $L_1 L_2 = M^2$. In such a case no nodal admittance matrix exists, but instead the ratio of $V_1 - V_2$ to $V_3 - V_4$ is always equal to $\sqrt{(L_1/L_2)}$. An ideal element, the ideal transformer, may be derived by letting $L_1$ and $L_2$ become infinite, keeping their ratio a constant, $n^2$. The ideal transformer is defined, in the notation of Fig. 5, by the equations

$$V_1 - V_2 = n(V_3 - V_4)$$

$$I_1 = -I_3/n$$

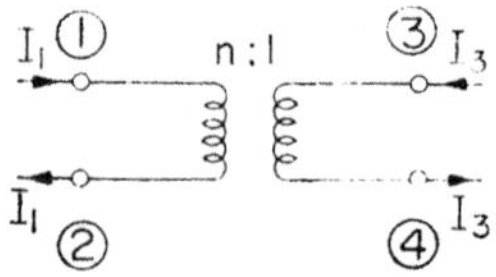


FIG. 5. *Ideal transformer*

It therefore establishes equations similar in nature to the equations describing network constraints and if these constraints are not accounted for explicitly, the set of nodal equations will prove singular since redundant variables are contained in them.

The network of Fig. 5 is still reciprocal in the sense that

$$\sum_{r=1}^{4} V_r' I_r'' = \sum_{r=1}^{4} V_r'' I_r'$$

but it will readily be shown that both sums are identically zero.

## GENERALISATION OF MUTUAL INDUCTANCE COUPLING

There is no need to restrict the number of windings involved to two, as considered earlier. In general a number of windings may be coupled inductively. The generalisation to a set of coils of $n$

windings may be simply made. If the notation of Fig. 6 is adopted the consideration of magnetic coupling gives the equations

$$\left.\begin{aligned}\hat{V}_1 &= pL_1 I_1 + pM_{12} I_2 + \ldots + pM_{1n} I_n \\ \hat{V}_2 &= pM_{21} I_1 + pL_2 I_2 + \ldots + pM_{2n} I_n \\ \\ \hat{V}_n &= pM_{n1} I_1 + pM_{n2} I_2 + \ldots + pL_n I_n\end{aligned}\right\} \quad (15)$$

The voltages here are not nodal voltages but the voltages across the individual windings. It may be shown from the properties of the magnetic field that $M_{rs} = M_{sr}$. Providing equations (15) can be inverted an exact parallel to the case of two windings may be followed. It may be proved that inversion is possible if no two windings are perfectly coupled. Because equations (15) are symmetrical, the inverted equations will also yield a symmetrical form. Paralleling the network with any *RLC* network will again

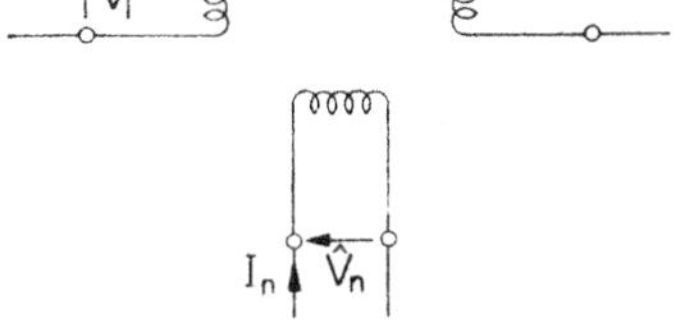


FIG. 6. *Multi-winding transformer*

give a symmetrical nodal admittance matrix. A network containing mutual inductance may be termed an *RLCM* network.

*Examples on Chapter 2*

1. Write down the nodal admittance matrices for each of the two networks below, taking node 4 as reference.

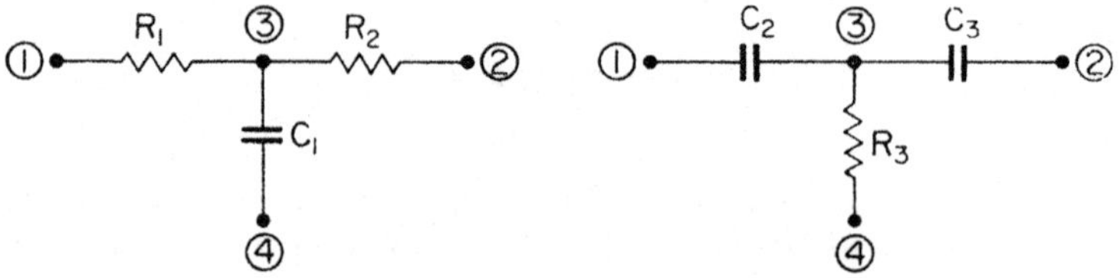


*Answer:*

$$\begin{bmatrix} G_1 & 0 & -G_1 \\ 0 & G_2 & -G_2 \\ -G_1 & -G_2 & G_1+G_2+pC_1 \end{bmatrix} \quad \begin{bmatrix} pC_2 & 0 & -pC_2 \\ 0 & pC_3 & -pC_3 \\ -pC_2 & -pC_3 & pC_2+pC_3+G_3 \end{bmatrix}$$

2. Rewrite these with new numbering as shown below, allowing for an unconnected node in each network. Hence obtain the nodal admittance matrix for the twin-T network which results when nodes of the same number are joined together.

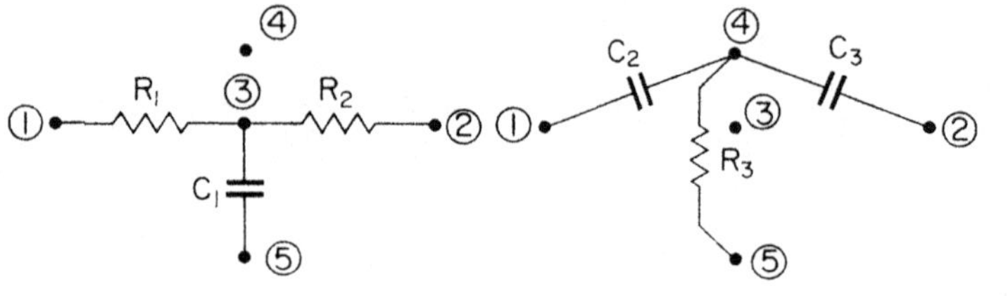


*Answer:*

$$\begin{bmatrix} G_1+pC_2 & 0 & -G_1 & -pC_2 \\ 0 & G_2+pC_3 & -G_2 & -pC_3 \\ -G_1 & -G_2 & G_1+G_2+pC_1 & 0 \\ -pC_2 & -pC_3 & 0 & pC_2+pC_3+G_3 \end{bmatrix}$$

3. In the three-winding mutually coupled circuit below the inductances of the coils are all equal to $L$, and the mutual inductance of any one coil to either of the others is $M$. Show that $I_1+I_2+I_3$ is zero, and find the nodal admittance matrix taking node 3 as reference.

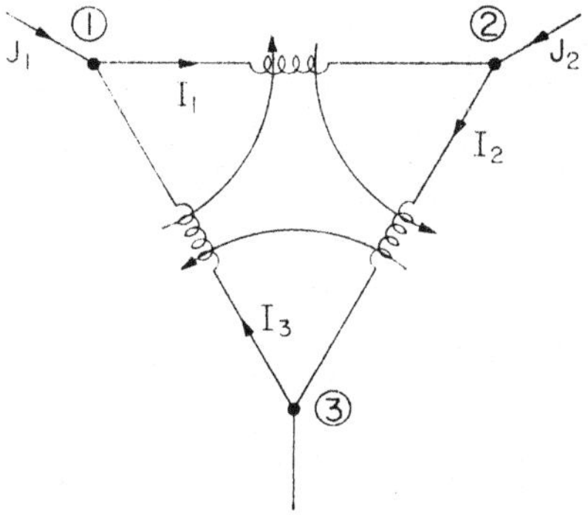


*Answer:*

$$\begin{bmatrix} \dfrac{2}{p(L-M)} & -\dfrac{1}{p(L-M)} \\ -\dfrac{1}{p(L-M)} & \dfrac{2}{p(L-M)} \end{bmatrix}$$

# 3 CLASSIFICATION AND DESCRIPTION OF CERTAIN NETWORKS

## PASSIVE NETWORKS

A passive network is one in which none of the components, other than voltage or current sources, is capable of generating electrical energy. Evidently the type of network previously considered comes in this category.

## ACTIVE NETWORKS

Any network in which the above condition is violated may be termed an active network. A component generating energy must have available a source from which to provide this energy but which is not considered to be part of the network. Obvious examples are networks involving valves and transistors. It is possible to design circuits using such devices operating linearly with small signals which have the properties of negative resistances, inductances or capacities. The criterion of an active network is that for some electrical conditions it must be possible to obtain, or dissipate, more power than is available from the sources which are considered as driving the network. In order that this shall be possible, types of component other than the passive elements of resistance, inductance and capacity must be introduced. One possibility lies in allowing these elements to occur with negative values. A negative resistance will generate energy instead of dissipating it. Similarly negative inductances will deliver energy to a network as the current increases instead of storing it. These elements do not however enable equivalent circuits to be produced for the linear operation of valves and transistors. A further class must be introduced which may be termed controlled generators. Such elements must be regarded as having two pairs of terminals (which may however have one common member).

The useful elements in this category can be reduced to four. These have as output terminals either a voltage source or a current source, and the input terminals are driven either by an applied voltage or by a circulating current. The magnitude of the source is proportional to the driving voltage or current. It is also only necessary to consider the input terminals as presenting an infinite impedance when they respond to a voltage source and zero impedance when they respond to a circulating current.

Each of these elements as so defined takes no input power, but provides output power in the same way as an ordinary voltage or current source.

## RECIPROCAL NETWORKS

A reciprocal network is one satisfying the reciprocity conditions considered earlier: these may be stated as symmetry of the coefficients $P_{rs}$ and $P_{sr}$ in the nodal admittance matrix, or the satisfaction of the terminal relation

$$\sum_{r=1}^{n-1} V_r' J_r'' = \sum_{r=1}^{n-1} V_r'' J_r'$$

It is immediately evident that any network made up of two terminal linear devices is reciprocal, whether these devices are passive $(R, L, C)$ or active elements. Further the addition of mutually coupled inductances does not destroy reciprocity.

It is possible to define a component which whilst passive is also non-reciprocal. Consider a two-terminal pair device for which the voltage across one pair of terminals is $v_1$ and that across the other is $v_2$, together with currents $i_1$, $i_2$ directed in the same way with respect to their voltages. The condition for energy balance is

$$v_1 i_1 + v_2 i_2 = 0$$

or

$$v_1 i_1 = -v_2 i_2$$

Two obvious sets of linear relationships between the voltages and currents which satisfy this equation are

$$\left.\begin{aligned} v_1 &= n v_2 \\ i_1 &= -\frac{1}{n} i_2 \end{aligned}\right\}$$

$$\frac{v_1}{i_2} = -\frac{v_2}{i_1} = s$$

The first set define an ideal transformer. The second set may be regarded as the defining equations of a linear passive device. This device has been given the name "gyrator". The gyrator is non-reciprocal. This may be seen by forming the sum

$$\begin{aligned} v_1' i_1'' + v_2' i_2'' &= s i_2' i_1'' - s i_1' i_2'' \\ &= -v_2'' i_2' - v_1'' i_1' \end{aligned}$$

The ideal transformer is, as seen above, reciprocal.

At the present time, a satisfactory practical passive element having the properties of a gyrator has not been realised. The concepts of reciprocity and non-reciprocity are however applicable to microwave circuits, and there the field equivalent of the gyrator has found great application.

## NON-RECIPROCAL NETWORKS

In general any circuit containing active elements such as valves or transistors (working in their linear range) will be non-reciprocal unless they are employed as two terminal elements.

Non-reciprocity is not however technically confined to active networks, since the gyrator as shown above is non-reciprocal whilst being a passive element. Practically speaking however, at the present time non-reciprocity in the field of linear lumped networks is confined to active networks.

## DESCRIPTION BY TERMINAL CONFIGURATION

The use of networks dictates a practical classification by the number of entries or exits through which electrical power may flow. These entries and exits are conveniently designated by the term "port". A different classification may be made by the number of terminals used. The "port" classification will be considered first.

## 1-PORT NETWORKS

These have two terminals only, and are hence classifiable as impedances (or admittances). The relation between the voltage across the terminals and the current may be deduced by an application of the solution of Chapters 1 and 2 for the general $n$-node network. In order to allow full generality, elements other than $RLC$ and $M$ must be allowed, although frequently we shall have occasion to restrict results to reciprocal networks only. The general solution is that of equation (8) of Chapter 1.

$$V_l = \sum_{k=1}^{n-1} \frac{\Delta_{lk}}{\Delta} I_k \tag{1}$$

The symbol for current has been changed to the more customary $I$. The elements of the determinant

$$\Delta = \begin{bmatrix} P_{11} & P_{12} & \cdot & \cdot & P_{1,n-1} \\ P_{21} & P_{22} & \cdot & \cdot & P_{2,n-1} \\ \cdot & \cdot & \cdot & \cdot & \cdot \\ P_{n-1,1} & P_{n-1,2} & \cdot & \cdot & P_{n-1,n-1} \end{bmatrix}$$

are now functions of the variable $p = j\omega$, and generally of the form $(1/pL + pC + G)$. They may in addition not satisfy the symmetry relation $P_{rs} = P_{sr}$.

If only two operating terminals are being considered (say node 1 and node $n$), equation (1) simplifies to

$$V_1 = \frac{\Delta_{11} I_1}{\Delta} \tag{2}$$

Hence the general form of an impedance function is

$$\frac{\Delta_{11}}{\Delta} \tag{3}$$

The properties of this expression as a function of $p$ will be considered in Chapter 8. For the present it is evident that one complex number describes the impedance at any single frequency.

## THEVENIN'S THEOREM

If only one node plus reference is being used but the other nodes may have injected into them constant currents, equation (1) may be written

$$V_1 = \frac{\Delta_{11}}{\Delta} I_1 + \sum_{k=2}^{n-1} \frac{\Delta_{1k}}{\Delta} I_k \tag{4}$$

This equation is of the form

$$V_1 = ZI_1 + V_0 \tag{5}$$

Equation (5) expresses Thevenin's theorem, that a given circuit containing constant generators and of which two terminals only are used may be regarded as a source of internal e.m.f. equal to the open-circuit voltage in series with an impedance.

The equivalent formulation in terms of a current generator is obtained by rearrangement of equation (5) and is called Norton's theorem.

## REACTIVE 1-PORT NETWORKS

If all the components are purely reactive, for real frequencies each term $P_{rs}$ in the nodal admittance matrix is a pure imaginary.

By the definition of a determinant of equation (9) Chapter 1, the ratio $\Delta_{11}/\Delta$ will in such a case be purely imaginary. This proves the obvious result that an impedance composed of reactances only is itself a reactance.

## 2-PORT NETWORKS

The next category is when two ports are used. In this usage external connection is made to the network through two pairs of terminals only, and current flows into one member of a pair and

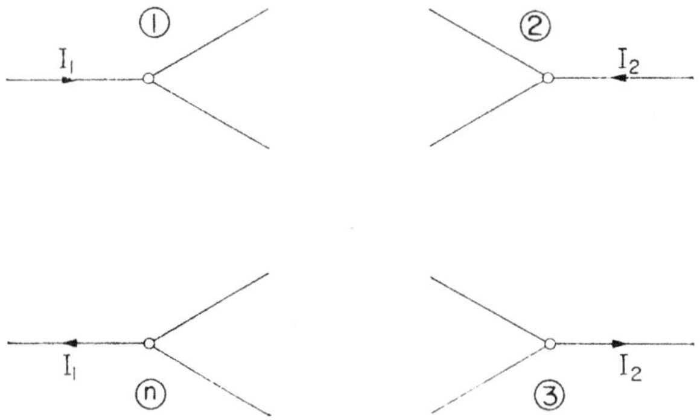


FIG. 1. *2-port connection*

out through the other. No current path exists between members of one pair and members of the other save internally in the network. This state of affairs is illustrated in Fig. 1. With this notation and regarding all other injected currents as zero, equations (1) reduce to

$$V_1 = \frac{\Delta_{11}}{\Delta} I_1 + \frac{\Delta_{12}}{\Delta} I_2 - \frac{\Delta_{13}}{\Delta} I_2$$

$$V_2 = \frac{\Delta_{21}}{\Delta} I_1 + \frac{\Delta_{22}}{\Delta} I_2 - \frac{\Delta_{23}}{\Delta} I_2$$

$$V_3 = \frac{\Delta_{31}}{\Delta} I_1 + \frac{\Delta_{32}}{\Delta} I_2 - \frac{\Delta_{33}}{\Delta} I_2$$

If the voltage $V_2 - V_3$ is denoted by $V_{23}$ we obtain

$$\left.\begin{aligned} V_1 &= \frac{\Delta_{11}}{\Delta} I_1 + \frac{\Delta_{12} - \Delta_{13}}{\Delta} I_2 \\ V_{23} &= \frac{\Delta_{21} - \Delta_{31}}{\Delta} I_1 + \frac{\Delta_{22} + \Delta_{33} - \Delta_{23} - \Delta_{32}}{\Delta} I_2 \end{aligned}\right\} \tag{6}$$

These may be rewritten in the form

$$\begin{aligned} V_1 &= z_{11} I_1 + z_{12} I_2 \\ V_{23} &= z_{21} I_1 + z_{22} I_2 \end{aligned} \tag{7}$$

Comparison of the coefficients gives expressions for $z_{11}$, $z_{12}$, $z_{21}$, $z_{22}$ in terms of $\Delta$ and its cofactors. If the network is reciprocal it follows that $z_{12} = z_{21}$. Hence, at a fixed frequency, four complex parameters are required to specify a 2-port network. If it is a reciprocal network this number reduces to three.

Incorporation of constant currents at the other nodes may readily be shown to give equations (7) with the addition of an e.m.f. in series with each of the leads to nodes 1 and 2. This provides a generalisation of Thevenin's theorem to 2-ports.

## MULTI-PORT NETWORKS

Apart from cases of coupled coils, multi-port networks are not of great interest in practical circuit analysis involving only lumped linear components. They are, however, of great interest in microwave circuits. The specification may be obtained by the same process as was adopted in deriving equations (6), and merely yields equations of the same form but with more terms.

## FURTHER DESCRIPTION OF 2-PORTS

In 2-port usage one port is used as an input and the second as an output. The effect of a generator between a terminal of one port and one in the other is not considered. Since no confusion will result in use the nomenclature will be changed slightly from that previously used and the symbols $V_1$ and $V_2$ used for the voltages

FIG. 2. *2-port network*

at the two ports and not for nodal voltages as hitherto. Referring to Fig. 2 and equations (7) we have

$$\left.\begin{aligned} V_1 &= z_{11}I_1 + z_{12}I_2 \\ V_2 &= z_{21}I_1 + z_{22}I_2 \end{aligned}\right\} \qquad (8)$$

The array

$$\begin{bmatrix} z_{11} & z_{12} \\ z_{21} & z_{22} \end{bmatrix}$$

is called the impedance matrix for the 2-port. In a network which is correctly described by the variables chosen, equations (8) may be used to express any pair of the four variables $V_1$, $V_2$, $I_1$, $I_2$ in terms of the remaining pair. Four further descriptions are then possible:

$$\left.\begin{aligned} I_1 &= y_{11}V_1 + y_{12}V_2 \\ I_2 &= y_{21}V_1 + y_{22}V_2 \end{aligned}\right\} \qquad (9)$$

$$\left.\begin{aligned} V_1 &= AV_2 - BI_2 \\ I_1 &= CV_2 - DI_2 \end{aligned}\right\} \tag{10}$$

$$\left.\begin{aligned} V_1 &= h_{11}I_1 + h_{12}V_2 \\ I_2 &= h_{21}I_1 + h_{22}V_2 \end{aligned}\right\} \tag{11}$$

$$\left.\begin{aligned} I_1 &= g_{11}V_1 + g_{12}I_2 \\ V_2 &= g_{21}V_1 + g_{22}I_2 \end{aligned}\right\} \tag{12}$$

The coefficients in each case can be expressed in terms of the coefficients of equations (8). The matrix

$$\begin{bmatrix} y_{11} & y_{12} \\ y_{21} & y_{22} \end{bmatrix}$$

is called the admittance matrix for the 2-port.

$$\begin{bmatrix} A & B \\ C & D \end{bmatrix}$$

may be called the cascade matrix. The reason for the negative signs in equations (10) is merely that, as will be seen, in a cascade

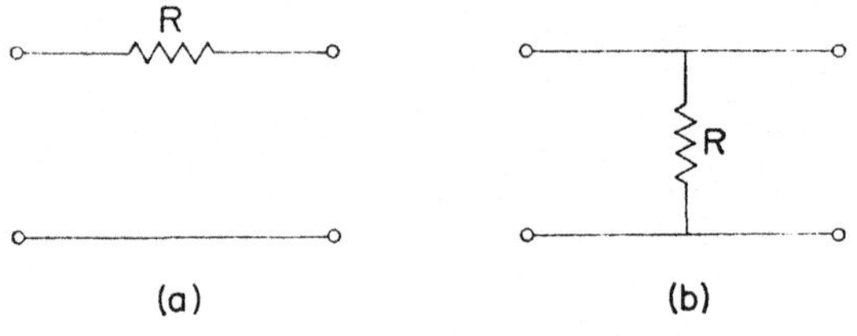


FIG. 3. *Simple 2-ports*

connection $-I_2$ is the output current, and is equal to the input current for the next unit.

The matrix formed by the coefficients of equations (11) is used in connection with transistor equivalent circuits, where the coefficients happen to be easily measured quantities.

The elements of any one of these matrices can be expressed in terms of those of any other, providing the equations are soluble. This is always so unless the original network has a peculiarity which means that it is overspecified by one set of equations. In such a case one set may exist whereas another set may not.

Examples of this state of affairs are easy to find: the ideal transformer imposes no relation between current and voltage. The only matrix description is the cascade matrix which becomes, if $n$ is the turns ratio,

$$\begin{bmatrix} n & 0 \\ 0 & 1/n \end{bmatrix} \tag{13}$$

The circuit of Fig. 3(*a*) has no impedance matrix representation and that of Fig. 3(*b*) no admittance matrix representation.

The admittance and cascade matrices of the circuit of Fig. 3(*a*) are respectively

$$\begin{bmatrix} G & -G \\ -G & G \end{bmatrix} \tag{14}$$

$$\begin{bmatrix} 1 & R \\ 0 & 1 \end{bmatrix} \tag{15}$$

The first of these is singular, and the equations cannot be inverted to yield an impedance matrix.

The impedance and cascade matrices for the circuit of Fig. 3(*b*) are respectively

$$\begin{bmatrix} R & R \\ R & R \end{bmatrix} \tag{16}$$

$$\begin{bmatrix} 1 & 0 \\ G & 1 \end{bmatrix} \tag{17}$$

Active 2-port elements can also be described by the sets of parameters specified by equations (8)–(12). For example the triode valve of Fig. 4(*a*) has an equivalent circuit as in Fig. 4(*b*), for small increments in the voltages and currents. $g_a$, $\mu$, $g_m$ are respectively the anode slope conductance, amplification factor and

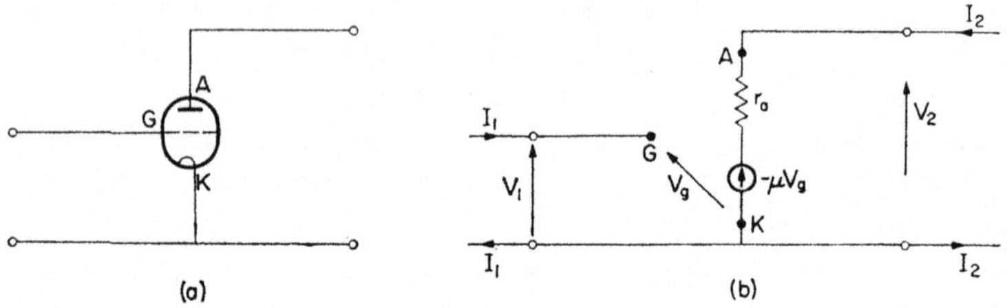


FIG. 4. *Non-reciprocal 2-port*

mutual conductance, satisfying the relation $g_m = g_a \mu$. The circuit of Fig. 4(*b*) has an admittance matrix

$$\begin{bmatrix} 0 & 0 \\ g_a & g_m \end{bmatrix} \tag{18}$$

and a cascade matrix

$$\begin{bmatrix} -\dfrac{1}{\mu} & \dfrac{1}{g_m} \\ 0 & 0 \end{bmatrix} \tag{19}$$

No impedance matrix exists.

By considering these circuits the meaning of the phrase used earlier, that the network is correctly described by the variables chosen, can be seen. Each of the circuits of Figs. 3 and 4 has some peculiarity which stops the network being described by any arbitrarily chosen pair of $V_1$, $V_2$, $I_1$, $I_2$. In Fig. 3(*a*) $I_1 = -I_2$ and hence they are not independent. In Fig. 3(*b*) $V_1 = V_2$ and in Fig. 4 $I_1 \equiv 0$.

## RELATIONS BETWEEN PARAMETERS OF A RECIPROCAL 2-PORT NETWORK

The particular relations existing in each set of parameters when reciprocity applies may be deduced by expressing them in terms of the elements of the impedance matrix, or by direct application of equation (11) of Chapter 1. Take the particular pair of experiments such that in the first $V_1' = 0$, and in the second $V_2'' = 0$. Equation (11) of Chapter 1 then gives

$$V_2' I_2'' = V_1'' I_1'$$

Applying this to equations (8)–(12) we have

$$z_{12} = z_{21} \tag{20}$$

$$y_{21} = y_{12} \tag{21}$$

$$AD - BC = 1 \tag{22}$$

$$h_{12} = -h_{21} \tag{23}$$

$$g_{12} = -g_{21} \tag{24}$$

## INTERCONNECTION OF 2-PORT NETWORKS

The interconnection of 2-port networks is a matter of considerable importance. The most important is that of cascading.

## CASCADE CONNECTION

Consider the circuit of Fig. 5
The cascade matrix expression of equation (10) gives

$$\begin{aligned} V_1 &= A_1 V_2 + B_1(-I_2) \\ I_1 &= C_1 V_2 + D_1(-I_2) \\ V_2 &= A_2 V_3 + B_2(-I_3) \\ -I_2 &= C_2 V_3 + D_2(-I_3) \end{aligned}$$

From these four equations $V_2$ and $I_2$ may be eliminated, giving two relations between $V_1$, $I_1$, $V_3$ and $I_3$. These equations are

$$\begin{aligned} V_1 &= (A_1 A_2 + B_1 C_2)V_3 + (A_1 B_2 + B_1 D_2)(-I_3) \\ I_1 &= (C_1 A_2 + D_1 C_2)V_3 + (C_1 B_2 + D_1 D_2)(-I_3) \end{aligned}$$

The process whereby these new coefficients are achieved from the original coefficients is more neatly put using the rule for the multiplication of matrices: if two square matrices of the same

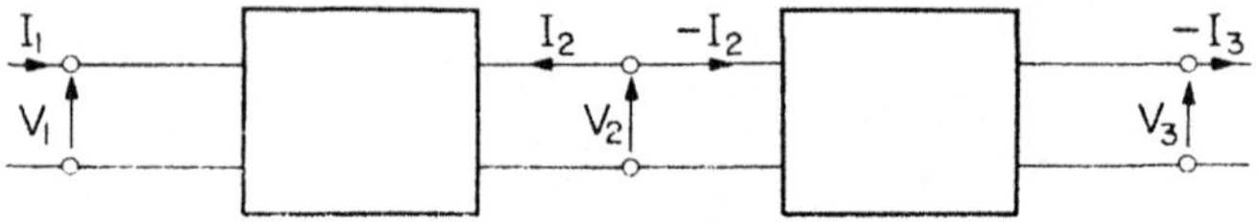


FIG. 5. *Cascade connection*

order, $n$, have elements $u_{ij}$, $v_{ij}$, the product is defined as the matrix having the element $w_{ij}$ where

$$w_{ij} = \sum_{k=1}^{n} u_{ik} v_{kj} \tag{25}$$

When $n = 2$ this may be explicitly written

$$\begin{bmatrix} w_{11} & w_{12} \\ w_{21} & w_{22} \end{bmatrix} = \begin{bmatrix} u_{11} v_{11} + u_{12} v_{21} & u_{11} v_{12} + u_{12} v_{22} \\ u_{21} v_{11} + u_{22} v_{21} & u_{21} v_{12} + u_{22} v_{22} \end{bmatrix} \tag{26}$$

Using this multiplication rule it is easily shown that

$$\begin{bmatrix} A_1 & B_1 \\ C_1 & D_1 \end{bmatrix} \times \begin{bmatrix} A_2 & B_2 \\ C_2 & D_2 \end{bmatrix} = \begin{bmatrix} A_1 A_2 + B_1 C_2 & A_1 B_2 + B_1 D_2 \\ C_1 A_2 + D_1 C_2 & C_1 B_2 + D_1 D_2 \end{bmatrix} \tag{27}$$

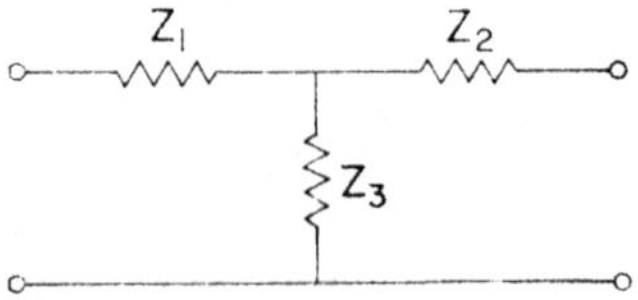


FIG. 6. *T-network*

This method enables the cascade matrix for a circuit to be evaluated in terms of the cascade matrices of simpler circuits.

As an example of this procedure consider the T-circuit of Fig. 6. This may be regarded as made up as shown in Fig. 7.

Using the expressions obtained earlier in equations (15) and (17) the cascade matrix of the 2-port of Fig. 6 may be written

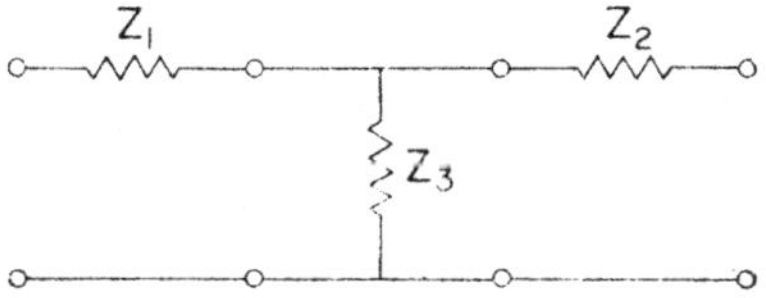


FIG. 7. *Decomposition of T-network*

$$\begin{bmatrix} 1 & Z_1 \\ 0 & 1 \end{bmatrix} \times \begin{bmatrix} 1 & 0 \\ 1/Z_3 & 1 \end{bmatrix} \times \begin{bmatrix} 1 & Z_2 \\ 0 & 1 \end{bmatrix}$$

This may be evaluated to give

$$\begin{bmatrix} 1+\dfrac{Z_1}{Z_3} & \dfrac{Z_1 Z_2 + Z_2 Z_3 + Z_3 Z_1}{Z_3} \\ \dfrac{1}{Z_3} & 1+\dfrac{Z_2}{Z_3} \end{bmatrix}$$

## PARALLEL CONNECTION

The parallel connection is shown in Fig. 8. Superficially, addition of the two sets of equation (9), in terms of the admittance matrix, gives

$$I_1 = (y'_{11}+y''_{11})V_1 + (y'_{12}+y''_{12})V_2$$
$$I_2 = (y'_{21}+y''_{21})V_1 + (y'_{22}+y''_{22})V_2$$

i.e. the admittance matrix of the combination is the sum of the individual admittance matrices. This result is correct, provided that the distribution of currents is as indicated in the figure. This is not, however, true for any pair of networks, such as for example those of Fig. 9. It is evident that the short circuit in the lower arm of the top network will profoundly modify the current distribution in the lower one. On the other hand, if both had been T-networks, the addition would have been correct.

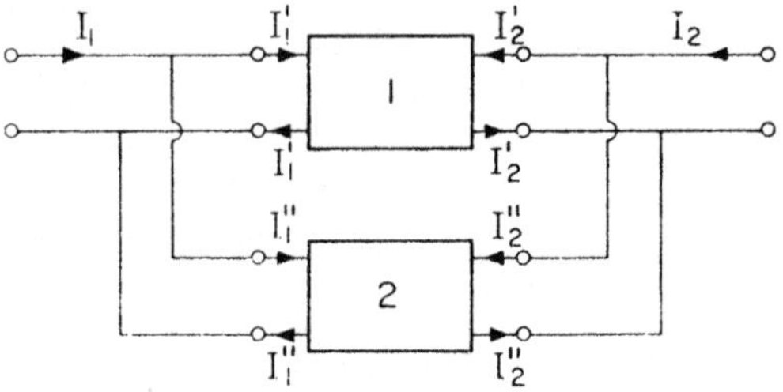


FIG. 8. *Paralleling of 2-ports*

In general the paralleling of networks violates the 2-port usage, and must be done using a complete nodal description involving the three terminals + reference terminal.

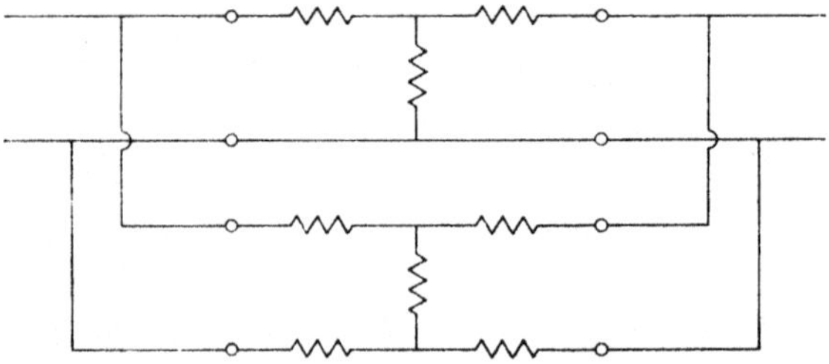

FIG. 9. *Networks which do not parallel as 2-ports*

Although the device is useful it must obviously be used with caution. It may be shown that if each end is paralleled in turn, and that at the opposite end no voltage exists between the terminals to be joined, the addition of admittance matrices is valid.

## SERIES INTERCONNECTION

This is shown in Fig. 10, and provided the current distribution satisfies the 2-port usage, addition of the two impedance matrices

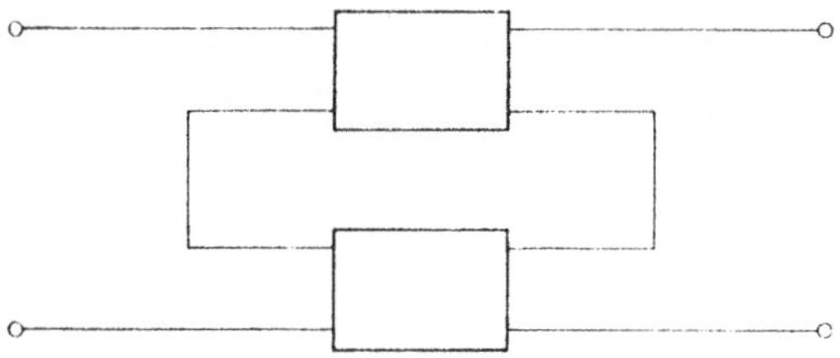

FIG. 10. *Series connection of 2-ports*

gives the resultant impedance matrix. The same condition for validity applies.

## REACTIVE 2-PORTS

2-port networks consisting of reactances only are important in the theory of filters. The nature of the describing parameters will now be considered. Any impedance associated with a purely reactive network must be a pure reactance. This automatically ensures that $z_{11}$ and $z_{22}$ of the impedance matrix are purely imaginary. It can also be shown that the transfer impedance $z_{21}$ is likewise reactive: consider the describing equations of a general reactive network, which are equations (12) of Chapter 2 with the special condition that all the coefficients $P_{rs}$ are imaginary, and equal to $jP'_{rs}$, $P'_{rs}$ real. If we have only one non-zero exciting current, $J_1$, the equation may be written

$$\sum_{s=1}^{n-1} P'_{rs} V_s = \begin{cases} -jJ_1 & v = 1 \\ 0 & v = 2, 3, \ldots, n-1 \end{cases}$$

Regarding $-jJ_1$ as an algebraic variable, on solution each $V_s$ will be a real number multiplied by $(-jJ_1)$, i.e. each $V_s$ has a phase angle of $\pm\pi/2$ with $J_1$. It also follows that the ratio of any two voltages will be $\pm 1$.

If we now apply this to a 2-port with $I_2 = 0$, the coefficient $z_{21}$ is seen to be purely imaginary. Similarly $z_{12}$ is imaginary.

Hence all the elements of the impedance matrix are imaginary, and a similar statement is true for the admittance matrix.

In the cascade matrix $A$ and $D$ are real, $B$ and $C$ purely imaginary.

## $m$-TERMINAL + REFERENCE NETWORKS

If of the $n$ nodes of a network, only $m$ ($< n$) are used for external connection, in addition to a reference node, the network may then be described as an $m$-terminal network + reference network, which will be abbreviated to $m$-terminal network. The equations relating the currents and voltages at the $m$ terminals may be deduced from the general equations for the $n$ node network. In order to avoid confusion it is desirable to regard the $m$ terminals as joined to the $m$ nodes through connections of zero resistance. This avoids ambiguity about current at the node and its relation to terminal current. If the $m$ terminals are taken as connected to nodes $1, 2, \ldots, m$, and nodes $m+1, m+2, \ldots, n-1$ are left unconnected, the describing equations in terms of nodal voltages become

$$\left.\begin{array}{llll}P_{11}V_1+ & \cdot\ \cdot & +P_{1,n-1}V_{n-1} & =I_1+I_{10}\\ \cdot & \cdot\ \cdot & \cdot & \cdot\\ P_{m1}V_1+ & \cdot\ \cdot & +P_{m,n-1}V_{n-1} & =I_m+I_{m0}\\ P_{m+1,1}V_1+ & \cdot\ \cdot & +P_{m+1,n-1}V_{n-1} & =I_{m+1,0}\\ \cdot & \cdot\ \cdot & \cdot & \cdot\\ P_{n-1,1}V_1+ & \cdot\ \cdot & +P_{n-1,n-1}V_{n-1} & =I_{n-1,0}\end{array}\right\} \tag{28}$$

In these equations the terms $I_1, I_2, \ldots, I_m$ refer to the currents into the terminals. The terms $I_{10}, \ldots, I_{n-1,0}$ refer to currents from internal current generators. When the reference node is connected (i.e. $\Delta \neq 0$) the first $m$ of equations (1) applied to equations (28) give

$$V_l = \sum_{k=1}^{m} \frac{\Delta_{lk}}{\Delta} I_k + \sum_{k=1}^{n-1} \frac{\Delta_{lk}}{\Delta} I_{k0}, \quad l = 1, 2, \ldots, m \tag{29}$$

Changing the notation these may be written in the form

$$V_l = \hat{V}_l + \sum_{k=1}^{m} L_{lk} I_k, \qquad l = 1, 2, \ldots, m \tag{30}$$

The $m \times m$ array of elements $L_{lk}$ is called the terminal impedance matrix describing the $m$-terminal network.

If the determinant of the elements $L_{lk}$ is formed, and if $K_{rs}$ denote the ratio of the cofactor of $L_{rs}$ in this determinant to the determinant itself, equations (30) may be inverted to give

$$\begin{aligned} I_r &= \sum_{s=1}^{m} K_{rs}(V_s - \hat{V}_s) \\ &= \hat{I}_r + \sum_{s=1}^{m} K_{rs} V_s \end{aligned} \tag{31}$$

The $m \times m$ array of the elements $K_{rs}$ is called the terminal admittance matrix of the $m$-terminal network.

If the reference node is not connected to the network, as may happen in particular cases, $\Delta = 0$ and equations (29) do not exist. In this case however equations (31) still exist. To prove this equations (28) must be used directly. The last $n-m-1$ of these equations are used to express $V_{m+1}, V_{m+2}, \ldots, V_{n-1}$ in terms of the other voltages and the currents. $V_{m+1}, \ldots, V_{n-1}$ can then be eliminated from the first $m$ equations, resulting in a set of equations of the form of equations (31). The array of elements $K_{rs}$ would then be called the indefinite terminal admittance matrix. It has the property that

$$\sum_{s=1}^{m} K_{rs} = 0$$

since in the absence of a connected reference node the total current entering the $m$ terminals must be zero. This procedure may still be followed when equations (29) do exist, giving an alternative derivation for the terminal admittance matrix.

## GENERALISATION OF THEVENIN'S AND NORTON'S THEOREMS TO THE $m$-TERMINAL NETWORK

Equations (30) and (31) provide immediate generalisations of these two theorems, at least when the reference node is connected.

If in equations (30) all terminals are open-circuited we find

$$V_r = \hat{V}_r$$

We may therefore replace the network by a network with the same terminal impedance matrix and with no internal generators but with voltage generators in each terminal lead given by $\hat{V}_r$. This is the generalisation of Thevenin's theorem. To obtain the same result for Norton's theorem, short-circuit all the terminals to the reference and measure the current in each lead. From equations (31)

$$I_r = \hat{I}_r$$

Hence the internal generators may be replaced by current generators of magnitude $\hat{I}_r$ directed outwards from each terminal node.

The extension to the indefinite case gives the same result for Norton's theorem. It also gives the result for Thevenin's theorem under extra conditions: that the representation by voltage generators is only possible if

$$\sum_{r=1}^{m} I_r = \sum_{r=1}^{m} \hat{I}_r = 0$$

This is physically obvious. Further, the voltage generators are arbitrary to a constant. This again follows physically from the arbitrariness of the reference.

## *Examples on Chapter 3*

1. Using the networks of Chapter 2 question 1, check the impedance between nodes 1 and 4 with node 2 disconnected, as given by the formula of equation (2).

2. In the twin-T network considered in Chapter 2 question 2 take $G_1 = G_2 = \frac{1}{2}G_3$, $C_2 = C_3 = \frac{1}{2}C_1$. Using the result for the nodal admittance matrix obtain an expression for the terminal admittance matrix when nodes 1, 2 are taken as the terminals with node 5 the common terminal. Show that $K_{12}$ vanishes at a real frequency.

*Answer:*
$$K_{11} = K_{22} = \frac{G_1^2 + pC_2^2 + 4G_1 pC_2}{2(G_1 + pC_2)}$$

$$K_{12} = -\frac{G_1^2 + p^2C_2^2}{2(G_1 + pC_2)}$$

$$\omega = G_1/C_2$$

3. Obtain impedance, admittance and cascade matrices for the three following 2-port networks:

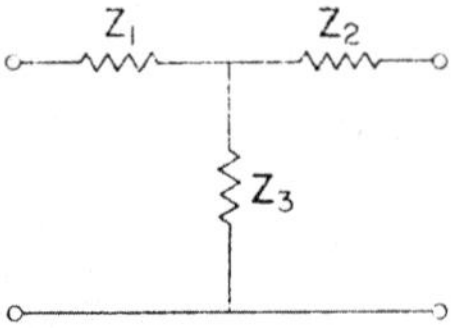


T–network

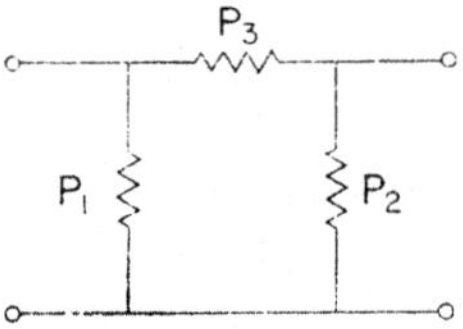


Π–network

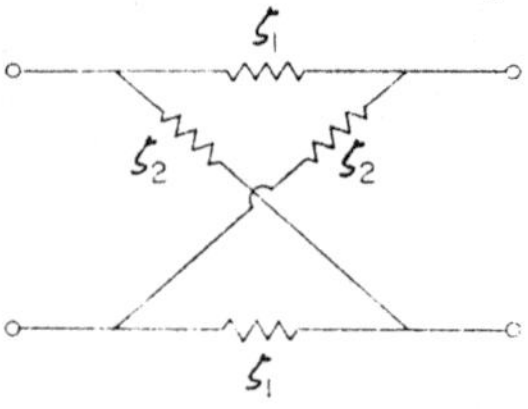


Lattice network

*Answer:*

| | Impedance Matrix | Admittance Matrix | Cascade Matrix |
|---|---|---|---|
| T | $\begin{bmatrix} Z_1+Z_3 & Z_3 \\ Z_3 & Z_2+Z_3 \end{bmatrix}$ | $\begin{bmatrix} \dfrac{Z_2+Z_3}{\Sigma Z_1Z_2} & -\dfrac{Z_3}{\Sigma Z_1Z_2} \\ -\dfrac{Z_3}{\Sigma Z_1Z_2} & \dfrac{Z_1+Z_3}{\Sigma Z_1Z_2} \end{bmatrix}$ | $\begin{bmatrix} 1+\dfrac{Z_1}{Z_3} & \dfrac{\Sigma Z_1Z_2}{Z_3} \\ \dfrac{1}{Z_3} & 1+\dfrac{Z_2}{Z_3} \end{bmatrix}$ |
| Π | $\begin{bmatrix} \dfrac{P_1(P_2+P_3)}{\Sigma P_1} & \dfrac{P_1P_2}{\Sigma P_1} \\ \dfrac{P_1P_2}{\Sigma P_1} & \dfrac{P_2(P_1+P_3)}{\Sigma P_1} \end{bmatrix}$ | $\begin{bmatrix} \dfrac{1}{P_1}+\dfrac{1}{P_3} & -\dfrac{1}{P_3} \\ -\dfrac{1}{P_3} & \dfrac{1}{P_2}+\dfrac{1}{P_3} \end{bmatrix}$ | $\begin{bmatrix} 1+\dfrac{P_3}{P_2} & P_3 \\ \dfrac{\Sigma P_1}{P_1P_2} & 1+\dfrac{P_3}{P_1} \end{bmatrix}$ |
| Lattice | $\begin{bmatrix} \frac{1}{2}(\zeta_2+\zeta_1) & \frac{1}{2}(\zeta_2-\zeta_1) \\ \frac{1}{2}(\zeta_2-\zeta_1) & \frac{1}{2}(\zeta_2+\zeta_1) \end{bmatrix}$ | $\begin{bmatrix} \dfrac{1}{2}\left(\dfrac{1}{\zeta_2}+\dfrac{1}{\zeta_1}\right) & \dfrac{1}{2}\left(\dfrac{1}{\zeta_2}-\dfrac{1}{\zeta_1}\right) \\ \dfrac{1}{2}\left(\dfrac{1}{\zeta_2}-\dfrac{1}{\zeta_1}\right) & \dfrac{1}{2}\left(\dfrac{1}{\zeta_2}+\dfrac{1}{\zeta_1}\right) \end{bmatrix}$ | $\begin{bmatrix} \dfrac{\zeta_2+\zeta_1}{\zeta_2-\zeta_1} & \dfrac{2\zeta_1\zeta_2}{\zeta_2-\zeta_1} \\ \dfrac{2}{\zeta_2-\zeta_1} & \dfrac{\zeta_2+\zeta_1}{\zeta_2-\zeta_1} \end{bmatrix}$ |

where

$$\Sigma Z_1Z_2 = Z_1Z_2+Z_2Z_3+Z_3Z_1$$

$$\Sigma P_1 = P_1+P_2+P_3$$

4. Form the cascade matrix for the network below. Find the frequency at which the phase of the voltage at end 2 differs by 180° from the voltage at end 1, end 2 being open-circuited. Find also the ratio of the magnitudes of the voltages under these conditions.

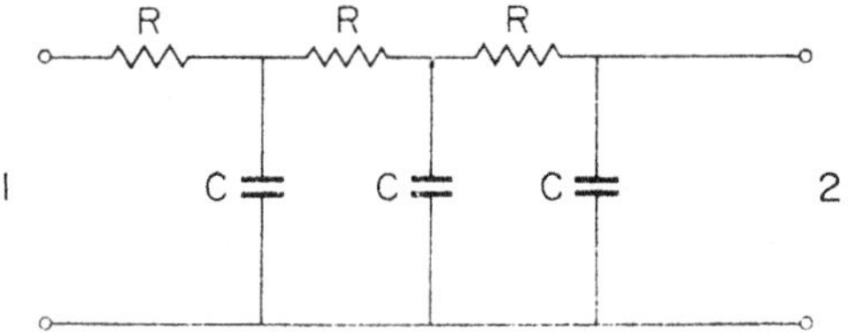


*Answer:*

$$\begin{bmatrix} 1+6x+5x^2+x^3 & R(3+4x+x^2) \\ pC(3+4x+x^2) & 1+3x+x^2 \end{bmatrix}$$

where

$$x = RCp$$

$$RC\omega = \sqrt{6},\ 29$$

5. Obtain an expression for the terminal admittance matrix, referred to node 3, for the network below.

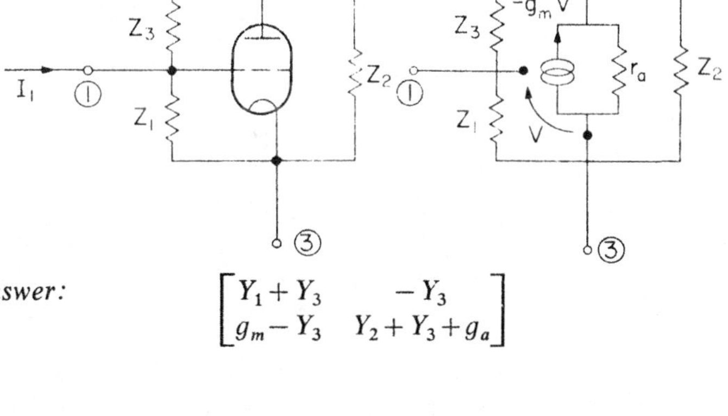


*Answer:*

$$\begin{bmatrix} Y_1+Y_3 & -Y_3 \\ g_m-Y_3 & Y_2+Y_3+g_a \end{bmatrix}$$

6. In the network of question 5 take the impedances $Z_1, Z_2, Z_3$ to be pure reactances of admittances $jB_1$, $jB_2$, $jB_3$ respectively. Show that if $I_2 = 0$ the admittance between nodes 1 and 3 will have a negative real part if

$$B_3^2(g_a + g_m) + g_m B_2 B_3 < 0$$

7. A 2-port with cascade matrix $\begin{bmatrix} A & B \\ C & D \end{bmatrix}$ is reversed in direction (i.e. ends 1, 2 are taken to be ends 2, 1 of a new 2-port). Find the cascade matrix with the new connection

*Answer:*
$$\frac{1}{AD-BC}\begin{bmatrix} D & B \\ C & A \end{bmatrix}$$

8. Regarding nodes 1, 2, 3 of the network below as the terminals of a 3-terminal network with reference node 5, find the terminal admittance matrix.

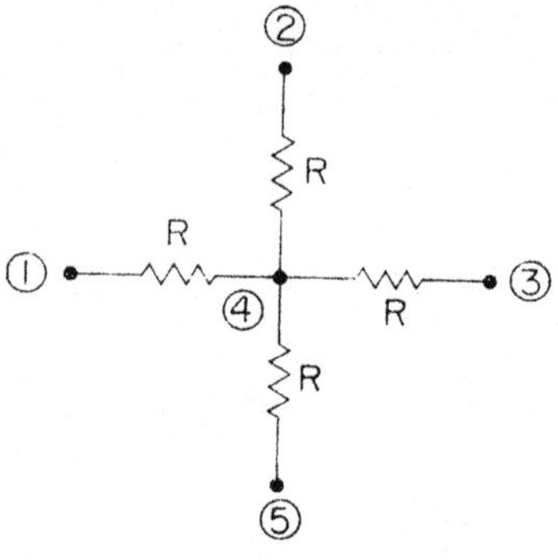


*Answer:*
$$\begin{bmatrix} \frac{3G}{4} & -\frac{G}{4} & -\frac{G}{4} \\ -\frac{G}{4} & \frac{3G}{4} & -\frac{G}{4} \\ -\frac{G}{4} & -\frac{G}{4} & \frac{3G}{4} \end{bmatrix}$$

# 4 EQUIVALENT NETWORKS

## EQUIVALENT CIRCUITS

Two circuits, of any complexity, are said to be equivalent for any purpose if the relationships between the particular voltages and currents chosen are identical. This equivalence may be said to be complete if the networks are both of the same number of nodes and have the same nodal admittance matrix. They are then equivalent in the sense that any prescribed currents injected at the nodes will yield the same nodal voltages. Two $m$-terminal plus reference networks will be equivalent if they have the same terminal admittance or impedance matrices.

Equivalence of 2-port networks will be ensured by identity of any of the describing matrices. A "port" network is a particular usage of a general network, in the sense that the available terminals are paired and the currents into the two members of each pair are equal and opposite. Hence if two networks with the same number of terminals and the same terminal admittance matrices are used as multi-ports, they will be equivalent as multi-ports. The converse is not true however.

*Example.* Consider the circuit of Fig. 1. This is a six-node network used as a three-terminal plus reference (node 4) network. Proceeding from first principles, we obtain three equations by considering the potentials at the two unused nodes:

$$V_1 - RI_1 = V_2 - RI_2 = r(I_1 + I_2) + R(I_1 + I_2 + I_3)$$

$$R(I_1 + I_2 + I_3) = V_3 - RI_3$$

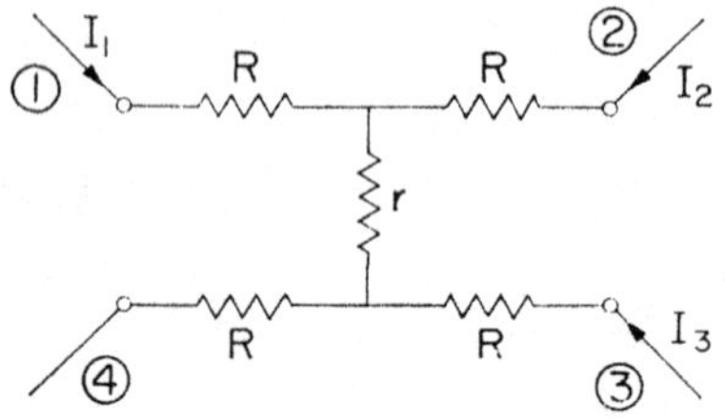


FIG. 1. *H-Network*

$V_1$, $V_2$, $V_3$ are voltages of the nodes relative to node 4. These equations may be rearranged to give

$$\left.\begin{aligned} V_1 &= (2R+r)I_1+(R+r)I_2+RI_3 \\ V_2 &= (R+r)I_1+(2R+r)I_2+RI_3 \\ V_3 &= RI_1 \qquad\quad +RI_2 \qquad\quad +2RI_3 \end{aligned}\right\} \tag{1}$$

The array

$$\begin{bmatrix} 2R+r & R+r & R \\ R+r & 2R+r & R \\ R & R & 2R \end{bmatrix}$$

is the terminal impedance matrix. These equations may be solved to give

$$I_1 = \frac{1}{4R^2(R+r)}\{(3R^2+2Rr)V_1-(R^2+2Rr)V_2-R^2V_3\}$$

$$I_2 = \frac{1}{4R^2(R+r)}\{-(R^2+2Rr)V_1+(3R^2+2Rr)V_2-R^2V_3\}$$

$$I_3 = \frac{1}{4R^2(R+r)}\{-R^2V_1-R^2V_2+(3R^2+2Rr)V_3\}$$

giving the terminal admittance matrix

$$\frac{1}{4R^2(R+r)}\begin{bmatrix} 3R^2+2Rr & -(R^2+2Rr) & -R^2 \\ -(R^2+2Rr) & 3R^2+2Rr & -R^2 \\ -R^2 & -R^2 & 3R^2+2Rr \end{bmatrix}$$

These elements may also be obtained directly from the definition (cf. equation (31) Chapter 3): connect nodes 2 and 3 to node 4 and inject $I_1$. Calculate $V_1$, $I_2$ and $I_3$. These results give the first column. The others may be similarly found.

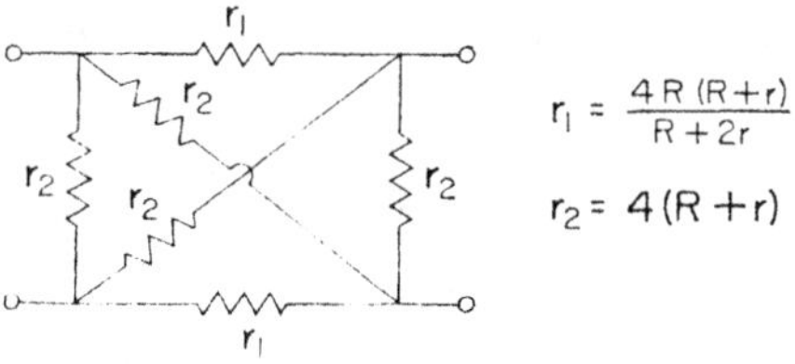


FIG. 2. *Equivalent circuit*

One possible equivalent circuit is obtained by regarding this terminal admittance matrix as the nodal admittance matrix of a four-node network. Using the result for the elements of the nodal admittance matrix that $P_{rs} = -$admittance between nodes $r$ and $s$, $P_{rr}$ = total admittance between node $r$ and all other nodes, the circuit of Fig. 2 is obtained.

It may be observed that although the circuit of Fig. 1 has resistance only, the equivalence will hold for arbitrary impedances in place of $R$ and $r$. The difference will be that the equivalent arms in Fig. 2 may not have simple representation in terms of $L$, $R$ and $C$.

## EQUIVALENCE OF 2-PORTS

Identity of any of the matrices describing the 2-port will ensure equivalence. Continuing the example of Fig. 1, when used as a

2-port with the ports between nodes 1 and 4, and nodes 2 and 3, equations (1) are modified by placing $I_3 = -I_2$. We find

$$V_1 = (2R+r)I_1 + rI_2$$

$$V_2 - V_3 = V_{23} = rI_1 + (2R+r)I_2$$

Hence the impedance matrix of the 2-port is

$$\begin{bmatrix} 2R+r & r \\ r & 2R+r \end{bmatrix}$$

This is also the impedance matrix of the circuit of Fig. 3. It is

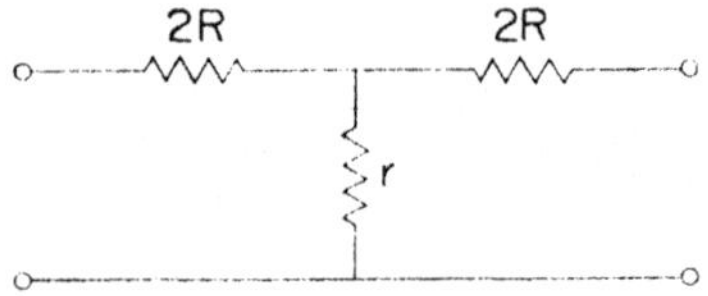


FIG. 3. *Equivalent circuit*

immediately evident that this is not equivalent to Fig. 1 as a three-terminal plus reference network. Regarded as a 2-port, the fact that the input current $I_1$ circulates round one side only means that any resistance in series can be placed in either arm without altering the 2-port properties. Practical reasons sometimes give a

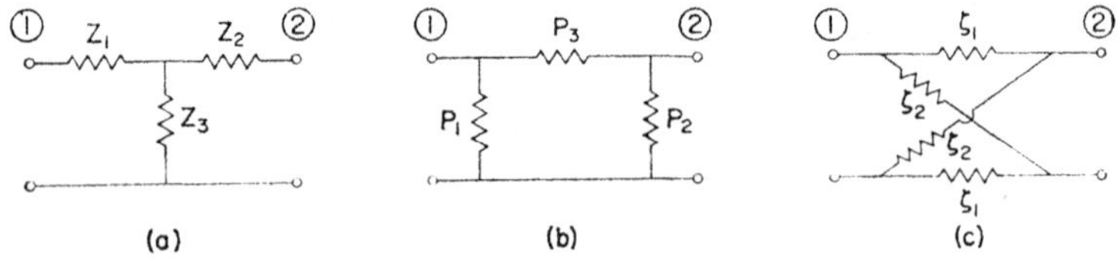


FIG. 4. *Common 2-port networks*

preference for use of one type rather than another. For example, the circuits of Figs. 1 and 3 may be attenuators. For most purposes that of Fig. 3 would be preferred as having an input terminal in common with one output terminal, but the H-type circuit of Fig. 1 could be preferable if the circuit was to be used symmetrically.

Some of the more common 2-port configurations will now be considered.

## COMMON 2-PORT NETWORKS

Three common types of 2-port are shown in Figs. 4(*a*), (*b*) and (*c*). These are referred to as the T, Π, and balanced lattice circuits. The symbols all represent impedances, the different ones being used for easy differentiation between the networks.

Conditions for equivalence between T- and Π-circuits may easily be obtained. The impedance matrices of the circuits of Fig. 4(*a*) and 4(*b*) are respectively

$$\begin{bmatrix} Z_1+Z_3 & Z_3 \\ Z_3 & Z_2+Z_3 \end{bmatrix}, \begin{bmatrix} \dfrac{P_1(P_3+P_2)}{P_1+P_2+P_3} & \dfrac{P_1 P_2}{P_1+P_2+P_3} \\[2ex] \dfrac{P_1 P_2}{P_1+P_2+P_3} & \dfrac{P_2(P_3+P_1)}{P_1+P_2+P_3} \end{bmatrix}$$

These expressions are most simply obtained by injecting $I_1$ into node 1, with node 2 disconnected, and vice versa. Hence the components of a T-circuit equivalent to the Π-circuit are

$$\left.\begin{aligned} Z_3 &= \frac{P_1 P_2}{P_1+P_2+P_3} \\ Z_1 &= \frac{P_1 P_3}{P_1+P_2+P_3} \\ Z_2 &= \frac{P_2 P_3}{P_1+P_2+P_3} \end{aligned}\right\} \tag{2}$$

The reverse process, expressing $P_1$, $P_2$ and $P_3$ in terms of $Z_1$,

$Z_2$ and $Z_3$ may be achieved by solution of equations (2), or more elegantly by forming the admittance matrix instead of the impedance matrix. To find the elements of this, short the output, i.e. $V_2 = 0$, and calculate $I_1 = y_{11} V_1$ and $I_2 = y_{21} V_1$. The calculation is then repeated for $V_1 = 0$. The results are

$$\begin{bmatrix} \dfrac{Z_2+Z_3}{\sum Z_1 Z_2} & \dfrac{Z_3}{\sum Z_1 Z_2} \\[2ex] \dfrac{Z_3}{\sum Z_1 Z_2} & \dfrac{Z_1+Z_3}{\sum Z_1 Z_2} \end{bmatrix}, \begin{bmatrix} \dfrac{P_1+P_3}{P_1 P_3} & \dfrac{1}{P_3} \\[2ex] \dfrac{1}{P_3} & \dfrac{P_2+P_3}{P_2 P_3} \end{bmatrix}$$

where $$\sum Z_1 Z_2 = Z_1 Z_2 + Z_2 Z_3 + Z_3 Z_1$$

Hence for equivalence

$$\left.\begin{aligned} P_3 &= \frac{\sum Z_1 Z_2}{Z_3} \\ P_1 &= \frac{\sum Z_1 Z_2}{Z_2} \\ P_2 &= \frac{\sum Z_1 Z_2}{Z_1} \end{aligned}\right\} \qquad (3)$$

It is to be noticed that for both these circuits the impedance matrix of the 2-port is also the terminal-impedance matrix of a three-terminal network, and hence the equivalence is complete as far as any connections to the three terminals are concerned.

The balanced lattice of Fig. 4($c$) has impedance and admittance matrices given respectively by

$$\begin{bmatrix} \frac{1}{2}(\zeta_2+\zeta_1) & \frac{1}{2}(\zeta_2-\zeta_1) \\ \frac{1}{2}(\zeta_2-\zeta_1) & \frac{1}{2}(\zeta_2+\zeta_1) \end{bmatrix}, \begin{bmatrix} \dfrac{1}{2}\left(\dfrac{1}{\zeta_1}+\dfrac{1}{\zeta_2}\right) & \dfrac{1}{2}\left(\dfrac{1}{\zeta_1}-\dfrac{1}{\zeta_2}\right) \\[2ex] \dfrac{1}{2}\left(\dfrac{1}{\zeta_1}-\dfrac{1}{\zeta_2}\right) & \dfrac{1}{2}\left(\dfrac{1}{\zeta_1}+\dfrac{1}{\zeta_2}\right) \end{bmatrix}$$

Hence equivalent T- and Π-circuits are as shown in Fig. 5($a$) and 5($b$).

Balanced latticed equivalents to T- and Π-circuits are only possible if the circuits are symmetrical. If $Z_1 = Z_2$ and $P_1 = P_2$ equivalent lattice circuits are given in Figs. 6(*a*) and 6(*b*). In these figures, and in later ones, the dotted lines indicate that the lattice is symmetrical and that the missing elements are identical to those in the corresponding arms shown.

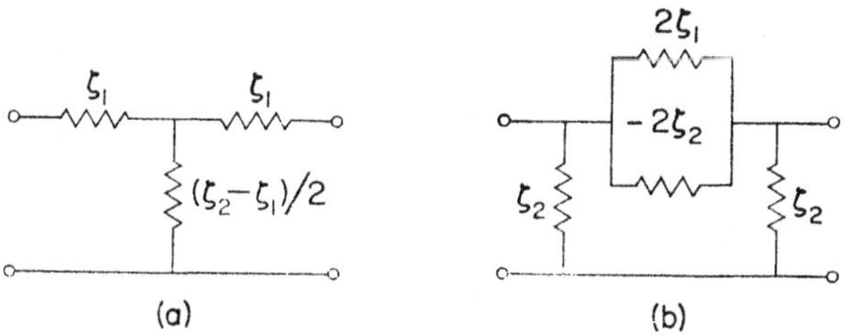


FIG. 5. *Networks equivalent to a balanced lattice*

It may be remarked that the circuits of Figs. 5(*a*) and 5(*b*) contain impedances which are the negative of real impedances. It is impossible with passive elements to produce negative inductance,

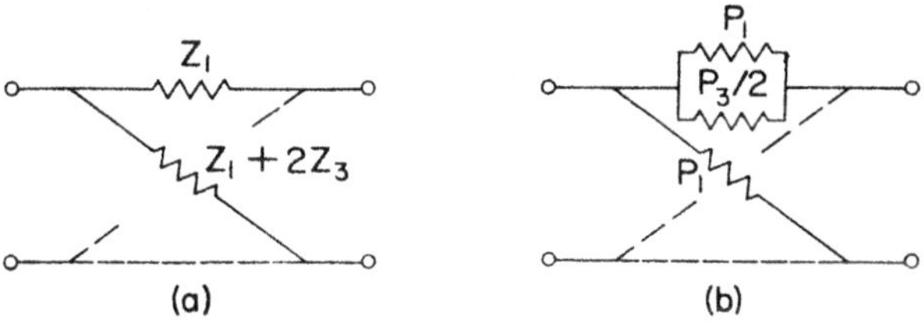


FIG. 6. *Lattice networks equivalent to T- and Π-networks*

capacitance or resistance. Accordingly these circuits are of practical, as opposed to theoretical, use only if $\zeta_1$ and $\zeta_2$ are such that the arms can be realised with passive elements. The lattice is in fact a more general type of network than the T- or Π-circuit.

Consideration could also be given to the unbalanced lattice, which is more general still. It is, however, very intractable to circuit transformations and has found little use.

## EQUIVALENT SOURCES

The problem of equivalent sources has been referred to in the previous chapter, and internal sources can be replaced by equivalent voltage or current generators in the leads. A somewhat different problem of much greater importance is that representation of active networks, or generally, non-reciprocal networks.

## EQUIVALENT CIRCUITS FOR ACTIVE NETWORKS

The general problem here is to produce a circuit containing passive elements and controlled generators equivalent to an active device such as a valve or transistor operating in its linear range. Consideration will be restricted here to 2-terminal plus reference circuits which cover the important 3-terminal devices of valves and transistors used as linear amplifiers.

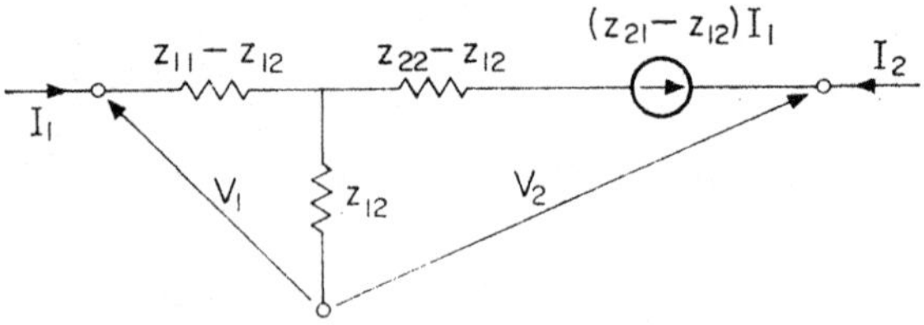


FIG. 7. *Network equivalent to a non-reciprocal network*

Complete description is of course given by any of the descriptive matrices, but these involve essentially working in 2-port matrix elements, which does not appear to be commonly acceptable. The only thing which stops the impedance matrix description of an active 2-port giving an equivalent T-circuit is the lack of reciprocity. Using the usual notation

$$V_1 = z_{11} I_1 + z_{12} I_2$$

$$V_2 = z_{21} I_1 + z_{22} I_2$$

We may write the second of these

$$V_2 = z_{12} I_1 + z_{22} I_2 + (z_{21} - z_{12}) I_1$$

An equivalent circuit is immediately given by Fig. 7. The lack of reciprocity has been accounted for by a voltage generator controlled by a current. This trick can obviously be done with any of the matrices and in either equation. Three more circuits are given in Figs. 8(*a*), (*b*) and (*c*).

In each of these circuits a controlled generator has been used. In some of these circuits a negative element may be involved. At low frequencies the internal electronic mechanism of active devices produces real coefficients in the impedance matrix, so that the negative elements will correspond to negative resistances, which are usually unacceptable in an equivalent circuit.

Further complications may be introduced by the desire to make the elements of the circuit bear a simple relation to measured quantities. This may lead to the incorporation of more than the one controlled generator which is strictly necessary.

Hence it will be seen that although it is simple to specify the necessary contents of such an equivalent circuit, the actual form chosen depends on many considerations, and a large number of equivalent circuits for the same device may find favour.

## CASCADE MATRICES FOR THE CONTROLLED GENERATORS

It may be noticed that the four controlled generators introduced in Chapter 3 have cascade matrices given by the following scheme:

Voltage output, voltage control $\begin{bmatrix} A & 0 \\ 0 & 0 \end{bmatrix}$

Voltage output, current control $\begin{bmatrix} 0 & 0 \\ C & 0 \end{bmatrix}$

Current output, voltage control $\begin{bmatrix} 0 & B \\ 0 & 0 \end{bmatrix}$

Current output, current control $\begin{bmatrix} 0 & 0 \\ 0 & D \end{bmatrix}$

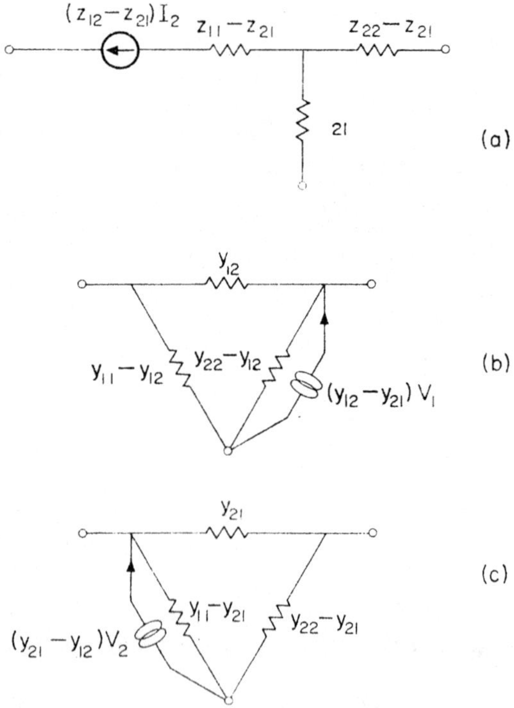


FIG. 8. *Networks equivalent to a non-reciprocal network*

*Examples on Chapter 4*

1. Obtain an equivalent circuit containing only three nodes plus reference for network of Chapter 3 question 8.

*Answer:*

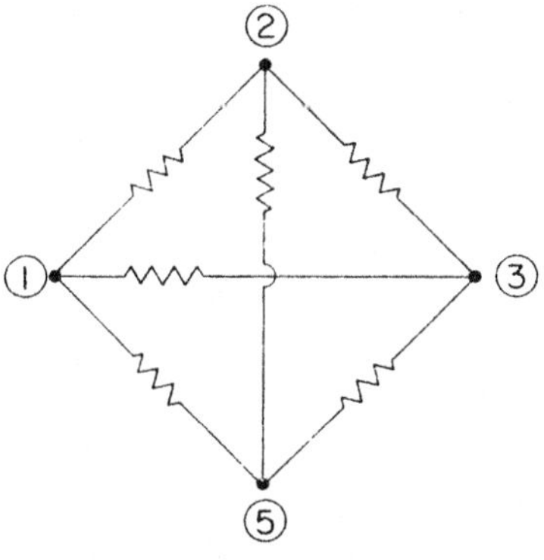


All resistors 4R

2. Equation (14) of Chapter 2 gives the nodal admittance matrix of the mutually coupled coils shown in Fig. 4 of that chapter. Obtain an equivalent circuit of four nodes only not involving mutual inductance.

*Answer:*

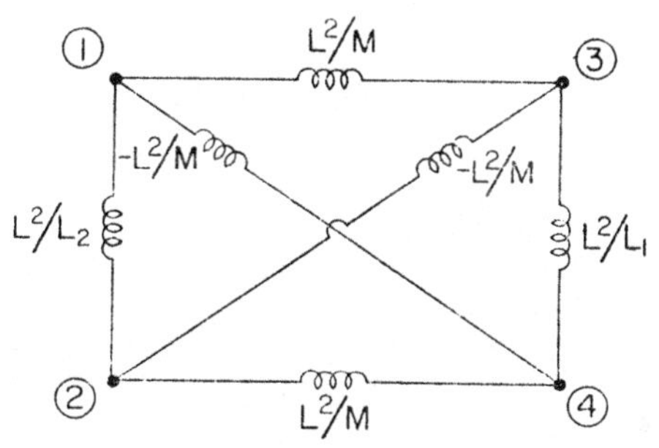


$L^2 = L_1 L_2 - M^2$

3. Find the 2-port impedance matrix of a pair of coupled inductors ($L_1$, $L_2$, $M$) and hence find a T-network equivalent for 2-port representation.

*Answer:*

$$\begin{bmatrix} pL_1 & pM \\ pM & pL_2 \end{bmatrix}$$

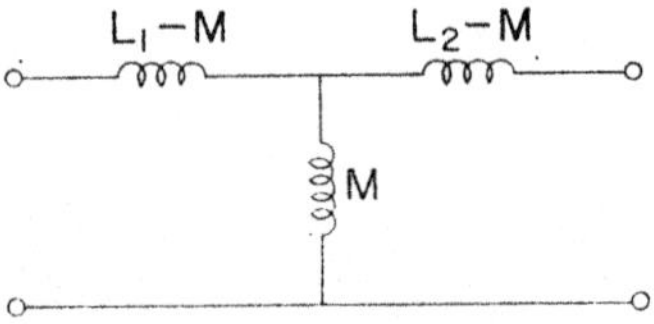


4. In the transformer networks below the windings are all of equal self inductance and equal mutual inductances occur between all pairs of coils. The current $I$ is provided from a current generator. Obtain a Π-equivalent network for the terminals 1, 2 and 3.

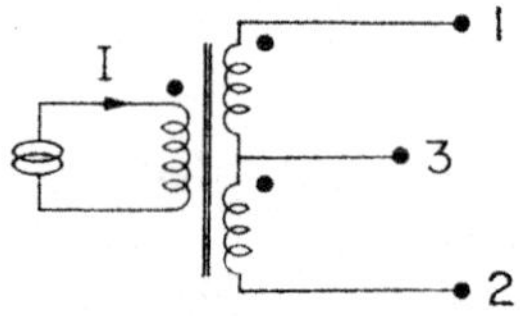


*Answer:*

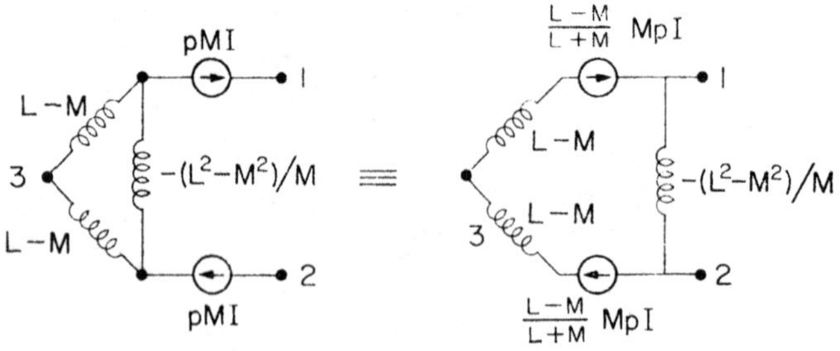

5. Show that the cascade matrix of the coupled inductances of question 3 is

$$\begin{bmatrix} \frac{L_1}{M} & p\frac{L_1 L_2 - M^2}{M} \\ \frac{1}{pM} & \frac{L_2}{M} \end{bmatrix}$$

6. Find the ratios of ideal transformers which when added after the mutual inductor of question 5 will give

(*a*) a symmetrical equivalent T-circuit;

(*b*) an equivalent circuit consisting of a series element followed by a shunt element.

*Answer:*

$$\sqrt{(L_2/L_1)}:1, \quad L_2/M:1$$

7. Obtain a balanced lattice equivalent to the network below for 2-port representation.

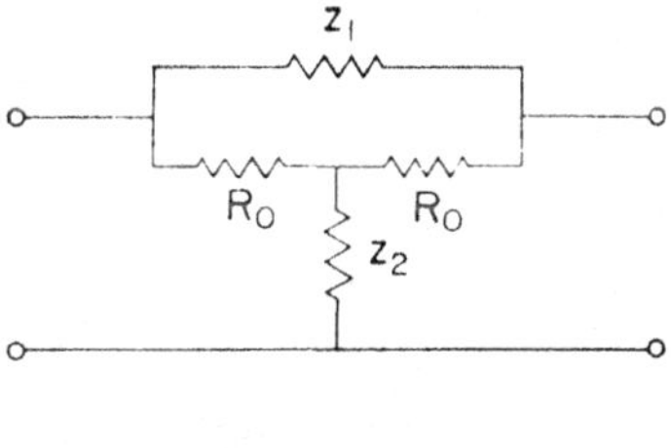


*Answer:* Series: $\frac{1}{2}Z_1$ in parallel with $R_0$

Shunt: $R_0$ in series with $2Z_2$.

8. Show that the networks below have 2-port representations equivalent to balanced lattices.

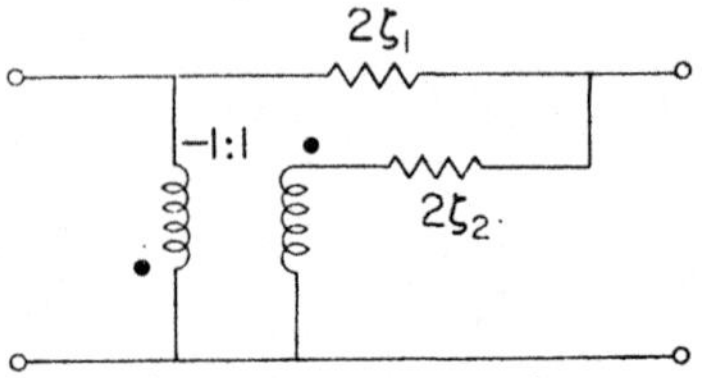


(Ideal transformer)

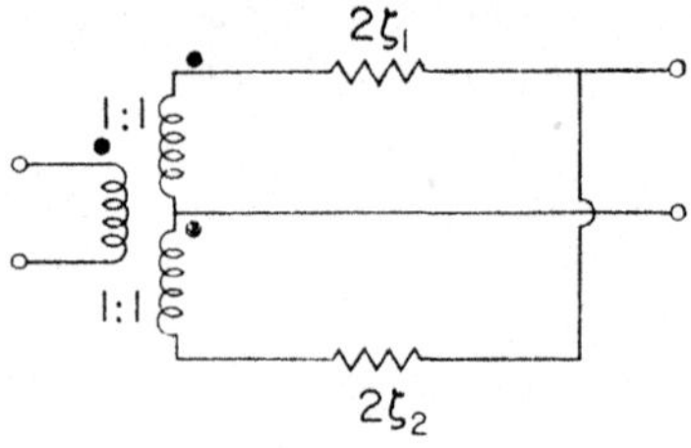


(Ideal three winding transformer)

9. Prove the identity of the 2-port representation of the balanced lattice networks below.

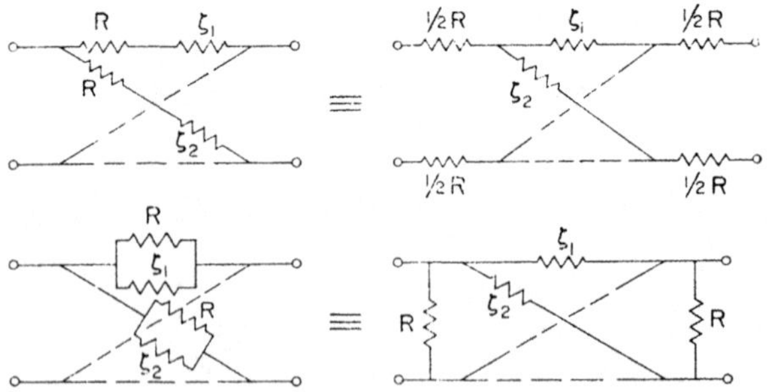

10. Demonstrate the equivalence of the following active 2-port networks:

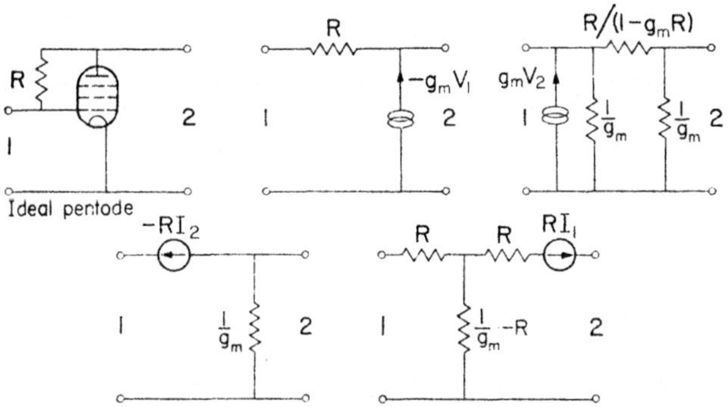

# 5 ITERATIVE AND IMAGE PARAMETERS OF 2-PORT NETWORKS

ONE of the special configurations in which 2-port networks are used is the cascade connection. Consideration of such a configuration leads to other methods of description particularly applicable to this case. The first one to be considered will be the iterative parameters.

## ITERATIVE PARAMETERS

These parameters are derived from the analysis of a cascade connection of a number of identical 2-port networks. An approach different from the straightforward analysis is now possible. This is to determine the restriction imposed upon the voltage between terminals by the cascade connection rather than by the source or load. Consider the circuit configuration of Fig. 1.

Using the cascade matrix

$$\left.\begin{aligned} V_{n-1} &= AV_n + BI_n \\ I_{n-1} &= CV_n + DI_n \\ V_n &= AV_{n+1} + BI_{n+1} \\ I_n &= CV_{n+1} + DI_{n+1} \end{aligned}\right\} \tag{1}$$

The first, third and fourth of these equations may be used to eliminate $I_n$ and $I_{n+1}$ giving the equation

$$(AD-BC)V_{n+1} + V_{n-1} = (A+D)V_n \tag{2}$$

Use of the second, third and fourth yields an identical equation between $I_{n-1}$, $I_n$ and $I_{n+1}$. This equation is a second-order difference equation. It may be shown that the most general solution contains two arbitrary constants. This particular equation may easily be found by substituting

$$V_n = \lambda\, \mathrm{e}^{n\Gamma} \tag{3}$$

The result is an equation in $\Gamma$:

$$(AD-BC)\,\mathrm{e}^{2\Gamma}+1 = (A+D)\,\mathrm{e}^{\Gamma}$$

Hence (3) is a solution of equation (2) if $\Gamma = \Gamma_1, \Gamma_2$ where

$$\left.\begin{aligned}\exp\Gamma_1, \exp\Gamma_2 & \\ &= \frac{1}{2(AD-BC)}[A+D\pm\{(A+D)^2-4(AD-BC)\}^{\frac{1}{2}}] \\ &= \frac{1}{2(AD-BC)}[A+D\pm\{(A-D)^2+4BC\}^{\frac{1}{2}}]\end{aligned}\right\} \tag{4}$$

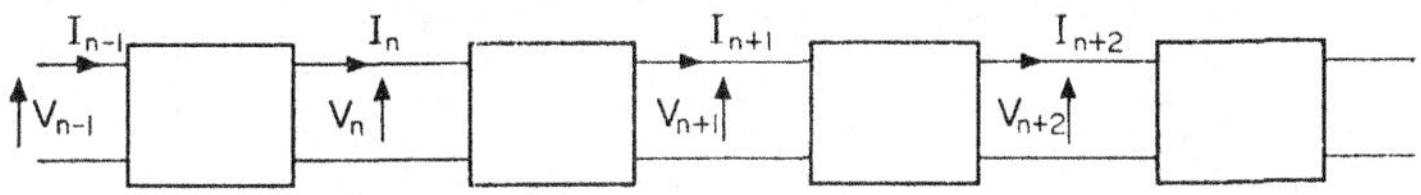


FIG. 1. *Cascade of 2-ports*

The existence of the square root makes it very difficult to say anything about $\Gamma_1$ and $\Gamma_2$ in an algebraic sense. One has however the relations

$$\exp(\Gamma_1+\Gamma_2) = (AD-BC)^{-1} \tag{5}$$

$$\exp\Gamma_1+\exp\Gamma_2 = \frac{A+D}{AD-BC} \tag{6}$$

If the network is reciprocal $AD-BC=1$ and we deduce from equation (5)

$$\Gamma_1 = -\Gamma_2 \tag{7}$$

A general expression for $V_n$ is given by

$$V_n = \lambda \exp(n\Gamma_1) + \mu \exp(n\Gamma_2) \tag{8}$$

in which $\lambda$, $\mu$ are arbitrary constants. Substitution into the first member of equations (1) gives for $I_n$ the expression

$$I_n = \frac{\lambda}{B}[\exp(-\Gamma_1) - A]\exp(n\Gamma_1) + \\ + \frac{\mu}{B}[\exp(-\Gamma_2) - A]\exp(n\Gamma_2) \tag{9}$$

$\Gamma_1$ and $\Gamma_2$ are called iterative propagation constants. Further discussion in general terms depends on algebraic consideration of the relation between the exponential terms and the matrix components, which is made difficult by the presence of the radical in equations (4). However, if we restrict ourselves to the passive reciprocal case, certain conclusions may be reached from other considerations. Initially let us consider networks which are not purely reactive. Equation (7) holds, and we define $\Gamma$ as the root in equations (4) which has a positive real part, which will always be possible for a lossy network. Equations (8) and (9) then take the form

$$V_n = \lambda e^{-n\Gamma} + \mu e^{n\Gamma} \tag{10}$$

$$I_n = \lambda \alpha_1 e^{-n\Gamma} + \mu \alpha_2 e^{n\Gamma} \tag{11}$$

where $\alpha_1$, $\alpha_2$ are defined by comparison with equation (9). If we consider a very long chain, the first term in both of these equations will decrease as $n$ increases positively, and the second will increase. Since we are dealing with a passive lossy network we must associate the first term with a source at $n=0$, say, and the second with a source at some large positive value of $n$. If only a source at $n=0$ is present, providing the chain is long enough, we may write

$$V_n = \lambda e^{-n\Gamma}$$

$$I_n = \lambda \alpha_1 e^{-n\Gamma}$$

for sufficiently small values of $n$. The impedance seen at the input to the chain is therefore $\alpha_1^{-1}$. This is also the impedance seen by the output of the first 2-port, looking into the chain. It is called an iterative impedance, and we may write

$$Z_{t1} = \alpha_1^{-1} = B(e^{\Gamma} - A)^{-1} \tag{12}$$

A second iterative impedance may be defined by considering a source at some large positive value of $n$. The approximate expressions for $V_n$ and $I_n$ for values of $n$ not too far from the source are then

$$V_n = \mu e^{n\Gamma}$$

$$I_n = \mu\alpha_2 e^{n\Gamma}$$

The impedance seen by the source is

$$Z_{t2} = -\alpha_2^{-1} = -B(e^{-\Gamma} - A)^{-1} \tag{13}$$

The negative sign arises because of the convention assumed for positive currents.

Using this notation for iterative impedances, equations (10) and (11) may be written

$$V_n = \lambda e^{-n\Gamma} + \mu e^{n\Gamma} \tag{14}$$

$$I_n = \frac{\lambda}{Z_{t1}} e^{-n\Gamma} - \frac{\mu}{Z_{t2}} e^{n\Gamma} \tag{15}$$

These equations may be given a "wave" interpretation, thinking of the first terms as corresponding to a forward travelling wave, and the second term as a backwards travelling wave. It will be noticed that the direction of the wave has been determined by the attenuation and not by phase change along the chain. If a purely reactive network is considered, for some frequencies at least $\Gamma$ may be purely imaginary. This makes it impossible to say that either of the terms of equations (10) and (11) will decrease with increasing $n$, so that this method of distinguishing is not applicable. An alternative way of defining the iterative impedances is to use the fact that if the first network is terminated by $Z_{t1}$, the input impedance is also $Z_{t1}$. Similarly, looking at the chain in the reverse direction, an impedance $Z_{t2}$ connected at the input to the first network will result in an impedance $Z_{t2}$ looking into the output end. Applying the first of these ways, the equation for $Z_{t1}$ is

$$Z_{t1} = \frac{AZ_{t1}+B}{CZ_{t1}+D} \tag{16}$$

Solving we find

$$Z_{t1} = \frac{1}{2C}[A-D \pm \{(A-D)^2+4BC\}^{\frac{1}{2}}] \tag{17}$$

We observe that two values of $Z_{t1}$ are obtained. A parallel method for $Z_{t2}$ yields

$$Z_{t2} = \frac{1}{2C}[D-A \pm \{(A-D)^2+4BC\}^{\frac{1}{2}}] \tag{18}$$

The significance of these double-valued expressions may be investigated by consideration of the lossy case. If, for lossy networks, we denote by $E$ that value of the radical $\{(A-D)^2+4BC\}^{\frac{1}{2}}$ which makes $\Gamma$ have a positive real part, we may write from equation (4)

$$\left.\begin{aligned} e^{\Gamma} &= \tfrac{1}{2}(A+D+E) \\ e^{-\Gamma} &= \tfrac{1}{2}(A+D-E) \end{aligned}\right\} \tag{19}$$

Equations (12) and (13) then give

$$\begin{aligned} Z_{t1} &= 2B(D-A+E)^{-1} \\ Z_{t2} &= 2B(A-D+E)^{-1} \end{aligned}$$

Rationalisation of these two expressions gives

$$Z_{t1} = \frac{1}{2C}(E-D+A) \tag{20}$$

$$Z_{t2} = \frac{1}{2C}(E+D-A)$$

With the same notation equations (17) and (18) are

$$Z_{t1} = \frac{1}{2C}(A - D \pm E) \tag{21}$$

$$Z_{t2} = \frac{1}{2C}(D - A \pm E)$$

Hence the calculation from equation (16) yields for $Z_{t1}$ either the true value of $Z_{t1}$ or $-Z_{t2}$. Similarly for $Z_{t2}$ we obtain either the true value of $Z_{t2}$ or $-Z_{t1}$. By reasoning from physical principles in the lossy case we have been enabled to distinguish between the two values obtained, and to single out a physical impedance which could be measured, in a particular case, to a very close degree of approximation and which is therefore realisable. In this lossy case $Z_{t1}$ and $Z_{t2}$ evidently contain positive resistance. Mathematically, termination at large $n$ in the impedance $-Z_{t2}$ would also give $-Z_{t2}$ at the input, but the impedance would not be realisable, since it would have a negative resistance.

We may alternatively distinguish between the two possible values for $Z_{t1}$ (i.e. $Z_{t1}$ or $-Z_{t2}$) obtained from equation (16) by applying some test of realisability. If they are resistive it is the same as before, that a negative resistance is unrealisable from passive elements. If purely reactive, we may use the criterion (proved later in Chapter 9 for finite networks) that the rate of change of reactance with frequency is proportional to the stored energy and is therefore positive.

It is relevant to take a brief look at the logic behind the development of iterative parameters. Equations (14) and (15) enable us to obtain the solution to the general problem of a chain of any number of identical networks with arbitrary terminations. The application of a condition at $n = N$, of termination by an impedance $Z_2$ gives a relation

$$V_N = Z_2 I_N$$

or

$$\lambda e^{-N\Gamma} + \mu e^{N\Gamma} = Z_2 \left( \frac{\lambda e^{-N\Gamma}}{Z_{t1}} - \frac{\mu}{Z_{t2}} e^{N\Gamma} \right)$$

Hence $$\frac{\lambda}{\mu}e^{-2N\Gamma} = \frac{(Z_2/Z_{t2})+1}{(Z_2/Z_{t1})-1} \tag{22}$$

An application of a further condition at $n = 0$, say a source of e.m.f. $E$ of internal impedance $Z_1$ gives

$$E = V_0 + Z_1 I_0$$

Hence $$E = \lambda + \mu + Z_1\left(\frac{\lambda}{Z_{t1}} - \frac{\mu}{Z_{t2}}\right) \tag{23}$$

Equations (22) and (23) enable $\lambda$ and $\mu$ to be determined. If $Z_2 = Z_{t1}$ we deduce immediately from (22) that $\mu = 0$. This is of course in agreement with previous results, giving a forward wave only. If the pseudo-value obtained for $Z_{t1}$, i.e. $-Z_{t2}$, is substituted for $Z_2$ we see that we have $\lambda = 0$, i.e. a backward wave only. Since $\Gamma$ was chosen to have a positive real part, this implies that with a source at $n = 0$ the voltage grows away from the source towards the termination, which can only be true if the termination is an active impedance.

In solving the practical case with two arbitrary terminating impedances, we make use of the results of two simple cases: those obtained with the particular termination $Z_{t1}$ and $Z_{t2}$ for forward and backward waves respectively. This leads to an alternative way of looking at the derivation of equations (14) and (15): an arbitrary terminating impedance can, for given voltage and current, be replaced by the impedance $Z_{t1}$ in series with an appropriate generator. Likewise a driving circuit can be replaced by a modified one of internal impedance $Z_{t2}$ and different e.m.f. Superposition of the two cases with sources at each end separately then gives equations (15). The fictitious generators are incorporated in the constants $\lambda$ and $\mu$ and have to be determined by applying boundary conditions.

This somewhat lengthy discussion has been included because some aspects of iterative parameters are frequently glossed over, and cause some confusion of thought. It may finally be emphasised that, for the purposes of the analysis of long chains of identical networks, the double-valued results obtained from equation (16) are quite irrelevant. The terms in equations (8) and (9) are quite unambiguous. Two values of $\Gamma$ are obtained, and these lead to unique coefficients in equation (9). It is only when a wave interpretation is placed on these equations that ambiguity enters.

Arguments of this type may be applied to the case of non-passive networks, but the matter is complicated by questions of stability.

## IMAGE PARAMETERS

The iterative method expounded is severely restricted since it deals with identical networks. Some relaxation to enable chains of different networks to be dealt with is desirable. Considering a single network in the chain of Fig. 1, the reasons for the success of the method are that two cases occur which have simple solutions. In these two cases the impedances presented at input and output of each network are the same: in the forward direction the impedance presented to the output of a network is always $Z_{t1}$ and this is the input impedance, and similarly for the backward direction.

If we can satisfy certain conditions we may expect to obtain solutions of the iterative type: these are roughly that in some ideal terminating condition a regular sequence of impedances at each junction occurs, the actual impedances to depend only on the networks adjacent to the junction.

One set of conditions which lead to suitable solutions is the following: in the ideal terminated condition (yet to be found) the impedance at a junction shall be the same in both directions (i.e. breaking the junction and looking into each circuit). Expressions for the necessary impedances for a given network can easily be found. Let them be $Z_{i1}$, $Z_{i2}$ on the input and output sides respectively. Then when terminated in $Z_{i2}$ the input impedance is $Z_{i1}$, and when $Z_{i1}$ is connected to the input, the output impedance is $Z_{i2}$. This state of affairs is illustrated in Fig. 2.

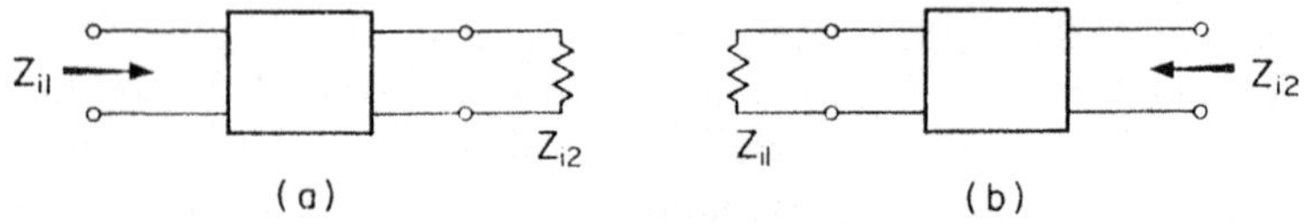


FIG. 2. *Connections defining image impedances*

For the circuits of Figs. 2(*a*) and 2(*b*) we have respectively

$$\left.\begin{aligned} Z_{i1} &= \frac{AZ_{i2}+B}{CZ_{i2}+D} \\ -Z_{i1} &= \frac{AZ_{i2}-B}{CZ_{i2}-D} \end{aligned}\right\} \tag{24}$$

Adding these two equations and solving we find

$$Z_{i2} = \left(\frac{BD}{AC}\right)^{\frac{1}{2}} \tag{25}$$

Substitution into the first equation, and rationalising the second yields

$$Z_{i1} = \frac{A}{D}Z_{i2} = \left(\frac{BA}{DC}\right)^{\frac{1}{2}} \tag{26}$$

The root in this equation is related for sign to that in equation (25). The first expression of equation (26) gives $Z_{i1}$ without ambiguity.

These impedances are termed the image impedances for the 2-port network. The relation between them is simpler than was the case with the two iterative impedances, and the choice of the appropriate expression to give a realisable impedance is much more simple.

The circumstances under which a chain may be set up to yield a simple solution require that we join networks which have the same image impedance at the junction. Such a chain is illustrated in the circuit of Fig. 3.

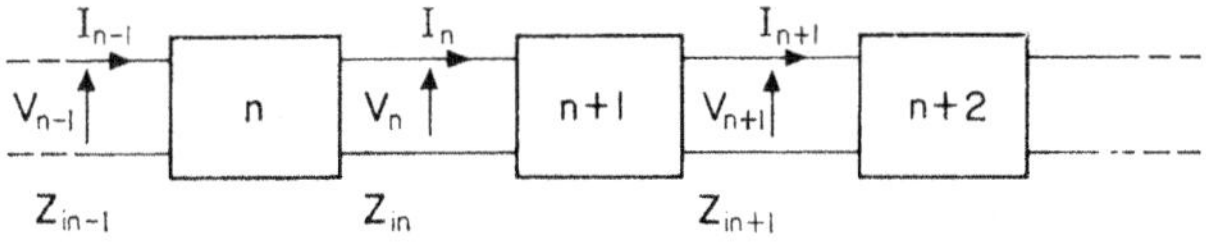


FIG. 3. *Image-matched 2-ports in cascade*

Network $n$ has input and output image impedances of $Z_{in-1}$, $Z_{in}$ respectively. Similarly network $n+1$ has input and output image impedances of $Z_{in}$ and $Z_{in+1}$. The equality of adjacent image impedances precludes using the method for a chain of arbitrary networks.

The two cases to be investigated are

(i) termination at some positive $n = N$ by $Z_{iN}$, source at $n = 0$,
(ii) termination at $n = 0$ in $Z_{i0}$, source at $N > 0$.

We may consider each network individually in these ideally terminated states, and the two relevant cases are illustrated in Fig. 4(*a*) and 4(*b*).

The equations for the circuit of Fig. 4(*a*) are

$$\left.\begin{aligned} I_n &= \frac{V_n}{Z_{in}} \\ I_{n-1} &= \frac{V_{n-1}}{Z_{in-1}} \\ V_{n-1} &= V_n\left(A_n + \frac{B_n}{Z_{in}}\right) \end{aligned}\right\} \tag{27}$$

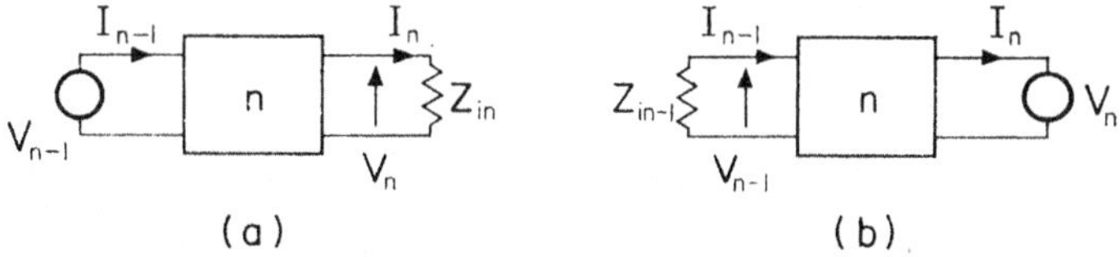


FIG. 4. *Image-matched 2-port circuits*

For the circuit of Fig. 4(*b*) we have

$$\left.\begin{aligned} I_n &= -\frac{V_n}{Z_{in}} \\ I_{n-1} &= -\frac{V_{n-1}}{Z_{in-1}} \\ V_{n-1} &= V_n\left(A_n - \frac{B_n}{Z_{in}}\right) \end{aligned}\right\} \tag{28}$$

Hence if we have the circuit of Fig. 3 terminated correctly at $n = N > 0$ and fed at $n = 0$ we may write

$$\left.\begin{aligned} V_n &= V_0 \prod_{r=1}^{n} (A_r + B_r/Z_{ir})^{-1} \\ I_n &= \frac{V_n}{Z_{in}} \end{aligned}\right\} \tag{29}$$

For the same circuit with $Z_{i0}$ at $n = 0$ and fed at $n = N$

$$\left.\begin{aligned} V_n &= V_0 \prod_{r=1}^{n} (A_r - B_r/Z_{ir})^{-1} \\ I_n &= -V_n/Z_{in} \end{aligned}\right\} \tag{30}$$

We may therefore write for a general solution of the circuit of Fig. 3.

$$\left.\begin{aligned} V_n &= \lambda \prod_{r=1}^{n}\left(A_r + \frac{B_r}{Z_{ir}}\right)^{-1} + \mu \prod_{r=1}^{n}\left(A_r - \frac{B_r}{Z_{ir}}\right)^{-1} \\ I_n &= \frac{\lambda}{Z_{in}} \prod_{r=1}^{n}\left(A_r + \frac{B_r}{Z_{ir}}\right)^{-1} - \frac{\mu}{Z_{in}} \prod_{r=1}^{n}\left(A_r - \frac{B_r}{Z_{ir}}\right)^{-1} \end{aligned}\right\} \tag{31}$$

The ambiguity in sign of $Z_{ir}$ is seen to be equivalent to interchange of $\lambda$ and $\mu$, since the change of sign of any one $Z_{ir}$ will involve change of sign of all in virtue of the matching conditions imposed.

In the case of a reciprocal network defined by these parameters only we would like to introduce one propagation constant to cover both product terms in equation (31). Consider the expression

$$\left(A_r-\frac{B_r}{Z_{ir}}\right)^{-1}=\frac{A_r+B_rZ_{ir}}{A_r^2-\left(\dfrac{B_r}{Z_{ir}}\right)^2}$$

$$=\frac{D_r}{A_r}\,\frac{A_r+B_rZ_{ir}}{A_rD_r-B_rC_r}$$

In the important case of reciprocal circuits we therefore have

$$\left[\left(\frac{D_r}{A_r}\right)^{\frac{1}{2}}\left(A_r-\frac{B_r}{Z_{ir}}\right)\right]^{-1}=\left[\left(\frac{D_r}{A_r}\right)^{\frac{1}{2}}\left(A_r+\frac{B_r}{Z_{ir}}\right)\right] \tag{32}$$

We therefore write

$$\left.\begin{aligned}\exp T_{1r}&=\left(\frac{D_r}{A_r}\right)^{\frac{1}{2}}\left(A_r+\frac{B_r}{Z_{ir}}\right)\\&=\left(\frac{Z_{ir}}{Z_{ir-1}}\right)^{\frac{1}{2}}\left(A_r+\frac{B_r}{Z_{ir}}\right)\\\exp T_{2r}&=\left(\frac{Z_{ir}}{Z_{ir-1}}\right)^{\frac{1}{2}}\left(A_r-\frac{B_r}{Z_{ir}}\right)\end{aligned}\right\} \tag{33}$$

$T_{1r}$ and $T_{2r}$ are the image propagation constants. If the network is reciprocal, equation (32) shows that

$$T_{1r}=-T_{2r}$$

The introduction of the square root introduces an ambiguity in $T$, but this causes no trouble if the term $(Z_{ir}/Z_{ir-1})^{\frac{1}{2}}$ is handled consistently.

Using the fact that adjacent image impedances are matched we may write equations (31) in the form

$$\left.\begin{aligned} V_n &= \sqrt{(Z_{in})}\left\{\lambda \exp\left(-\sum_{r=1}^{n} T_{1r}\right)+\mu \exp\left(-\sum_{r=1}^{n} T_{2r}\right)\right\} \\ I_n &= \sqrt{(1/Z_{in})}\left\{\lambda \exp\left(-\sum_{r=1}^{n} T_{1r}\right)-\mu \exp\left(-\sum_{r=1}^{n} T_{2r}\right)\right\} \end{aligned}\right\} \quad (34)$$

It is convenient to recapitulate the possible sources of ambiguity in these expressions. The term $\sqrt{Z_{in}}$ occurs in both equations, and ambiguity here has the same effect as changing the sign of both the arbitrary constants $\lambda$ and $\mu$. $Z_{in}$ itself is dual valued. If $Z_{in}$ in these equations is replaced by $-Z_{in}$, $\lambda$ and $\mu$ must be readjusted to include the imaginary multiplier, arising from the concealed $\sqrt{Z_{i0}}$, which has been tacitly incorporated. This will result in changing the sign throughout the second equation relative to the first. On the other hand the replacement in the equations (33) will interchange the actual values of $T_{1r}$ and $T_{2r}$ restoring the original equations. Thus, in any particular network, once an expression is taken as defining $Z_{in}$, $T_{1n}$ and $T_{2n}$ follow unambiguously. For any real termination of the networks the application of equations (34) will yield the correct answer, whichever value of $Z_{in}$ was chosen. If however a termination is $Z_{in}$, the result will obviously depend on which value is chosen.

## RECIPROCAL NETWORKS

The case of reciprocal networks is of great practical importance. In this case only one propagation constant is necessary, as previously shown. It is, however, necessary to establish a further convention: the term "image propagation constant" applied to a reciprocal network will be taken to mean the constant $T_{1r}$ of equations (33) with that value of $Z_{ir}$ which is a passive impedance. This does not imply that $Z_{ir}$ can be realised with a finite number of components, which is generally not possible. It is however the practically useful definition since approximations to this impedance can then be made with ordinary components. The appropriate value of $Z_{ir}$ can alternatively be defined as the one having a real positive part, or if reactive, for which the rate of change of reactance with frequency is positive.

## EXPRESSIONS FOR THE IMAGE PARAMETERS OF RECIPROCAL CIRCUITS IN TERMS OF OPEN- AND SHORT-CIRCUIT IMPEDANCES

The open- and short-circuit impedances for a 2-port network are defined as follows:

$Z_{o1}$ = impedance of end 1 when end 2 is open-circuited,
$Z_{s1}$ = impedance of end 1 when end 2 is short-circuited,
$Z_{o2}$ = impedance of end 2 when end 1 is open-circuited,
$Z_{s2}$ = impedance of end 2 when end 1 is short-circuited.

Expressions for these impedances may be easily obtained in terms of the elements of the cascade matrix. Putting in the conditions in brackets we have

$$\left.\begin{aligned} (I_2 = 0) \quad Z_{o1} &= A/C \\ (V_2 = 0) \quad Z_{s1} &= B/D \\ (I_1 = 0) \quad Z_{o2} &= D/C \\ (V_1 = 0) \quad Z_{s2} &= B/A \end{aligned}\right\} \tag{35}$$

We see immediately that

$$\frac{Z_{o1}}{Z_{s1}} = \frac{AD}{BC} = \frac{Z_{o2}}{Z_{s2}} = \frac{Z_o}{Z_s} \tag{36}$$

Further $$Z_{o1} Z_{s1} = \frac{AB}{CD} = Z_{i1}^2$$

$$Z_{o2} Z_{s2} = \frac{DB}{AC} = Z_{i2}^2$$

Hence
$$\left.\begin{aligned} Z_{i1} &= Z_{s1}\sqrt{(Z_o/Z_s)} = \sqrt{(Z_{s1}\,Z_{o1})} \\ Z_{i2} &= Z_{s2}\sqrt{(Z_o/Z_s)} = \sqrt{(Z_{s2}\,Z_{o2})} \end{aligned}\right\} \tag{37}$$

The middle expressions in these equations have been put in to define the image impedances in terms of the same square root, avoiding unnecessary ambiguity.

From equations (33), dropping the suffix 1 since we are restricted to reciprocal circuits,

$$e^T = (D/A)^{\frac{1}{2}}(A+B/Z_{i2})$$

As mentioned earlier the ambiguity introduced by the radical $(D/A)^{\frac{1}{2}}$ is irrelevant, since it is equivalent to changing the sign of both $\lambda$ and $\mu$ throughout equations (34). Using equations (37)

$$\begin{aligned} e^T &= \left(\frac{D}{A}\right)^{\frac{1}{2}}\left\{A+\frac{B}{Z_{s2}}\sqrt{\frac{Z_s}{Z_o}}\right\} \\ &= A\left(\frac{D}{A}\right)^{\frac{1}{2}}\left\{1+\sqrt{\frac{Z_s}{Z_o}}\right\} \end{aligned}$$

In a reciprocal circuit

$$AD-BC = 1$$

Hence
$$\frac{1}{AD} = 1-\frac{Z_s}{Z_o}$$

We may therefore write

$$\left.\begin{aligned} e^T &= \frac{1+\sqrt{(Z_s/Z_o)}}{(1-Z_s/Z_o)^{\frac{1}{2}}} \\ &= \left[\frac{1+\sqrt{(Z_s/Z_o)}}{1-\sqrt{(Z_s/Z_o)}}\right]^{\frac{1}{2}} \\ &= \left[\frac{\sqrt{(Z_o/Z_s)}+1}{\sqrt{(Z_o/Z_s)}-1}\right]^{\frac{1}{2}} \end{aligned}\right\} \tag{38}$$

We have also for this case $T_{2r} = -T_{1r} = -T$ (39)

*Examples on Chapter 5*

1. Set up equations (14) and (15) for the *R–C* network below in an approximate form valid for $(\omega RC)^{\frac{1}{2}} \ll 1$.

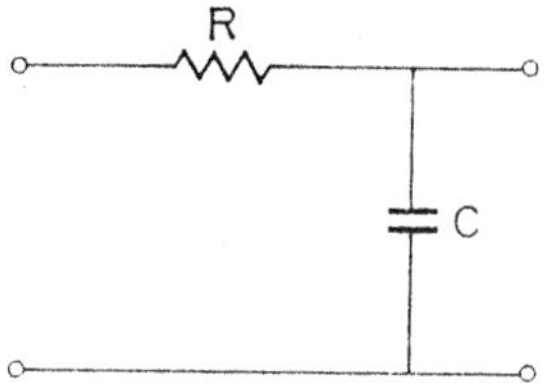


*Answer:*

$$\Gamma \simeq (\omega RC)^{\frac{1}{2}} \exp(j\pi/4)$$

$$\frac{1}{Z_{t1}} \simeq \frac{1}{Z_{t2}} \simeq \frac{1}{R}(\omega RC)^{\frac{1}{2}} \exp(j\pi/4)$$

2. A number of sections of the network of question 1 are joined together to form a chain which is open-circuited at the end $n = N$. Show that providing $(\omega RC)^{\frac{1}{2}} \ll 1$

$$V_N = V_0 \operatorname{sech} N\Gamma$$

If also $N \gg 1$ show that the lowest frequency for which $V_N$ is 180° out of phase with $V_0$ is given by

$$\omega RC \simeq 2\pi^2/N$$

and that

$$\left|\frac{V_0}{V_N}\right| = \cosh \pi$$

3. A feedback amplifier stage is shown in the circuit below. Obtain expressions for $\exp\Gamma_1$ and $\exp\Gamma_2$ as in equation (4). Show that for large values of $R_1$

$$\exp\Gamma_1 = \frac{G_2}{G_1} + \frac{2G_2 + g_m}{G_2} + \text{terms in } G_1$$

$$\exp \Gamma_2 = -\frac{g_m}{G_2} + (G_2 + g_m)^2 \frac{G_1}{G_2^3} + \text{higher powers of } G_1$$

$$\frac{1}{B}\{\exp(-\Gamma_1) - A\} = -G_2 - G_1\left(1 + \frac{g_m}{G_2}\right) + \ldots$$

$$\frac{1}{B}\{\exp(-\Gamma_2) - A\} = \frac{G_1(g_m + G_2)}{G_2} + \ldots$$

Consider the physical significance of the expressions by reference to a feedback amplifier consisting of several of these stages in cascade. It is convenient to take $g = 5 \times 10^{-3}$, $G_2 = 2 \times 10^{-4}$, $G_1 \simeq 10^{-6}$.

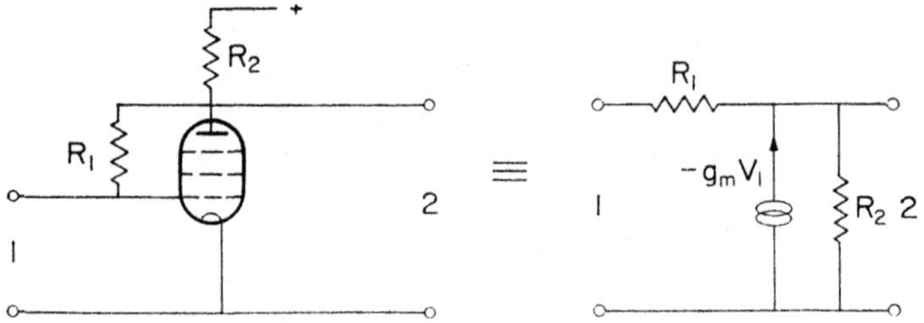


4. Show that the cascade matrix parameters may be expressed in terms of the image parameters according to the following relationships:

$$A = \tfrac{1}{2}\sqrt{(Z_{i1}/Z_{i2})}\{\exp T_1 + \exp T_2\}$$

$$B = \tfrac{1}{2}\sqrt{(Z_{i1}\, Z_{i2})}\{\exp T_1 - \exp T_2\}$$

$$C = \tfrac{1}{2}\sqrt{(1/Z_{i1}\, Z_{i2})}\{\exp T_1 - \exp T_2\}$$

$$D = \tfrac{1}{2}\sqrt{(Z_{i2}/Z_{i1})}\{\exp T_1 + \exp T_2\}$$

Hence show that $\begin{vmatrix} A & B \\ C & D \end{vmatrix} = \exp(T_1 + T_2)$

5. Show that the voltage along a chain of image matched networks as in Fig. 3 satisfies the difference equation

$$\begin{aligned} V_n Z_{in}^{-\frac{1}{2}}\{\exp(T_{1n}-T_{2,n+1})-\exp(T_{2n}-T_{1,n+1})\} \\ = V_{n-1} Z_{in-1}^{-\frac{1}{2}}\{\exp(-T_{2,n+1})-\exp(-T_{1,n+1})\}+ \\ +V_{n+1} Z_{in+1}^{-\frac{1}{2}}\{\exp T_{1n}-\exp T_{2n}\} \end{aligned}$$

Show that this is satisfied by equations (34).

6. A network for which the elements of the cascade matrix are all real is fed from a source of available power $P_0$, internal resistance $R_1$, and is terminated by a resistance $R_2$. Show that the maximum power transfer occurs for $R_1$ and $R_2$ equal to the respective image impedances, and that the ratio of output power to the input power is $P_0 \exp(-2T_1)$. [The e.m.f. of a source of available power $P_0$ is $2\sqrt{(R_1 P_0)}$.]

# 6 FILTER PROPERTIES OF REACTIVE 2-PORTS OPERATING BETWEEN RESISTIVE IMPEDANCES

THE term filter is used to describe a 2-port network designed to distinguish by its transmission properties different frequency ranges. For example a filter which passes with small attenuation all frequencies between zero and some upper limit and attenuates more greatly those above that limit is referred to as a low-pass filter. A characteristic of a typical low-pass filter is shown in Fig. 1.

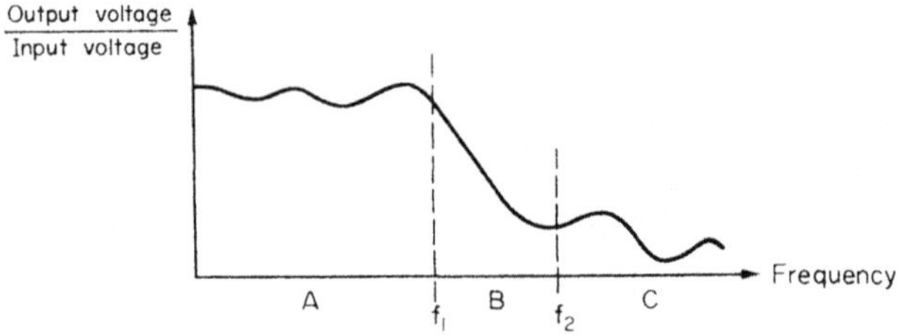


FIG. 1. *Typical response of a low-pass filter*

Three frequency ranges may be roughly designated. *A* may be called the pass band and *C* the stop band. These are separated by a range *B* which covers the change over of transmission characteristic from the pass to stop bands.

It must immediately be said that this definition is extremely loose. In any practical case it is usually necessary to specify as well as the frequency $f_1$ the permissible ratio of maximum to minimum output in the pass band (fixing the ripple), the maximum output in the stop band, and the frequency difference $f_2 - f_1$ which determines the rate of cut-off of the filter (and hence its ability to discriminate between neighbouring frequencies).

In principle there is no doubt that the correct approach is firstly to define a realisable network function which gives the required characteristic and then to synthesise a circuit which will give this characteristic response. This procedure is indeed used in modern filter design where the most stringent requirements are made. The systematic development of this procedure is however outside the scope of this book.

For many purposes an alternative design procedure is of great use, which is based on the properties of the image parameters of reactive 2-ports.

## THE IMAGE PARAMETERS OF REACTIVE 2-PORTS

Equation (38) of Chapter 5 gives the following expression for the image propagation constant $T$:

$$e^T = \left[\frac{\sqrt{(Z_o/Z_s)}+1}{\sqrt{(Z_o/Z_s)}-1}\right]^{\frac{1}{2}} \tag{1}$$

With a purely reactive 2-port the open- and short-circuit impedances must all be purely reactive, and hence the ratio $Z_o/Z_s$ is a real number. This ratio may be either positive or negative, and the two cases give two types of behaviour of the image parameters.

Let us denote

$$|Z_o/Z_s| = \sigma^2$$

(*i*) $Z_o/Z_s < 0$

$$\sqrt{(Z_o/Z_s)} = j\sigma \tag{2}$$

Hence
$$\left.\begin{aligned} Z_{i1} &= Z_{s1}\,j\sigma \\ Z_{i2} &= Z_{s2}\,j\sigma \end{aligned}\right\} \tag{3}$$

Since $Z_{s1}$, $Z_{s2}$ are reactive, these image impedances will be real, and the sign of the radical will be chosen to make them positive.

Further
$$e^{2T} = \frac{j\sigma+1}{j\sigma-1} \tag{4}$$

and
$$|e^{2T}| = 1$$

(*ii*) $Z_o/Z_s > 0$

$$\sqrt{(Z_o/Z_s)} = \sigma \tag{5}$$

Hence
$$\left.\begin{aligned} Z_{i1} &= Z_{s1}\,\sigma \\ Z_{i2} &= Z_{s2}\,\sigma \end{aligned}\right\} \tag{6}$$

Both these are now reactive.

$$e^{2T} = \frac{\sigma+1}{\sigma-1} \tag{7}$$

This is now purely real.

Generally we write

$$T = \alpha + j\beta$$

where $\alpha$ is the attenuation constant and $\beta$ is the phase constant.

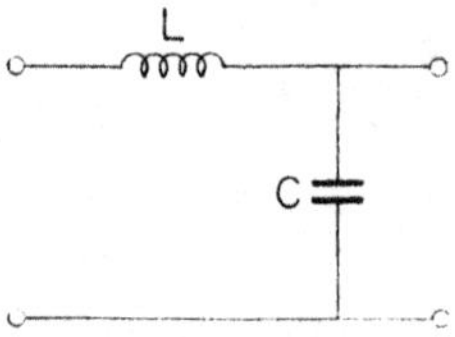


FIG. 2. *Illustrative circuit*

From the above we have

(*i*) $Z_o/Z_s < 0$, $\alpha = 0$, $\beta$ varies.

(*ii*) $Z_o/Z_s > 0$, $\alpha$ varies, $\beta$ constant.

*Example.* Consider the circuit of Fig. 2.

Here we have
$$Z_{o1} = j\omega L + 1/j\omega C$$
$$Z_{s1} = j\omega L$$
$$Z_{o2} = 1/j\omega C$$
$$Z_{s2} = (j\omega C + 1/j\omega L)^{-1}$$
$$\frac{Z_s}{Z_o} = \frac{\omega L}{\omega L - 1/\omega C} = \frac{1}{1-\omega_0^2/\omega^2}$$
where
$$\omega_0 = 1/\sqrt{(LC)}$$
Hence
$$0 < \omega < \omega_o \quad Z_s/Z_o < 0$$
$$\omega > \omega_o \quad Z_s/Z_o > 0$$
$$Z_{i1} = j\omega L(1-\omega_0^2/\omega^2)^{\frac{1}{2}}$$
$$Z_{i2} = \frac{j\omega L(1-\omega_0^2/\omega^2)^{\frac{1}{2}}}{1-\omega^2/\omega_0^2}$$

In order to get positive resistances for $Z_{i1}$, $Z_{i2}$ in the region $0 < \omega < \omega_0$ we must choose

$$\sqrt{(Z_o/Z_s)} = (1-\omega_0^2/\omega^2)^{\frac{1}{2}} = -j\sqrt{(\omega_0^2/\omega^2-1)} = j\sigma$$

the square root understood to be a positive quantity.

For $\omega > \omega_0$
$$\sqrt{(Z_o/Z_s)} = \sqrt{(1-\omega_0^2/\omega^2)} = \sigma$$

again understood as a positive quantity. This choice gives a positive slope to the reactances.

Using this convention, expressions for the image impedances are

$0 < \omega < \omega_0$:
$$Z_{i1} = \sqrt{(L/C)}\sqrt{(1-\omega^2/\omega_0^2)}$$
$$Z_{i2} = \sqrt{(L/C)}/\sqrt{(1-\omega^2/\omega_0^2)}$$

$\omega > \omega_0$:
$$Z_{i1} = j\sqrt{(L/C)}\sqrt{(\omega^2/\omega_0^2-1)}$$
$$Z_{i2} = -j\sqrt{(L/C)}/\sqrt{(\omega^2/\omega_0^2-1)}$$

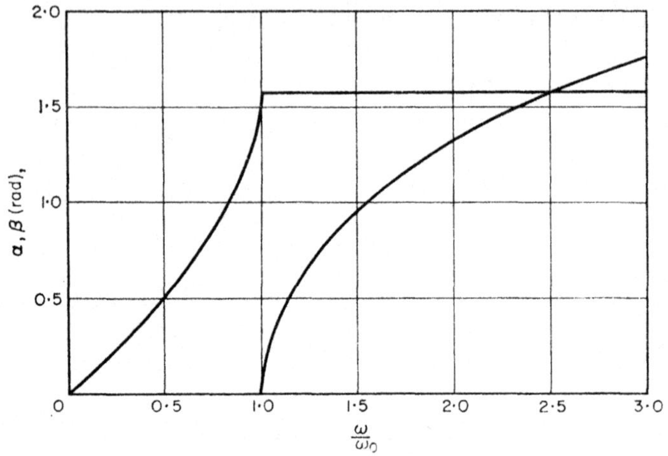


FIG. 3(*a*). *Attenuation and phase*

Expressions for the propagation constant may now be obtained:

$$0 < \omega < \omega_0: \quad e^{2T} = \frac{j\sigma+1}{j\sigma-1} = \frac{-j\sqrt{(\omega_0^2/\omega^2-1)}+1}{-j\sqrt{(\omega_0^2/\omega^2-1)}-1}$$

This will give $2T = 2j\beta$, but $2\beta$ will be arbitrary to $2\pi$. Since in this circuit $\omega = 0$ corresponds to straight-through connection, it is convenient to make this also $\beta = 0$.

By arrangement we see

$$e^{2T} = \frac{\sqrt{(1-\omega^2/\omega_0^2)}+j\omega/\omega_0}{\sqrt{(1-\omega^2/\omega_0^2)}-j\omega/\omega_0}$$

Hence if $\quad \sin\beta = \omega/\omega_0, \quad 0 \leqq \beta \leqq \pi/2$

$$T = j\beta$$

$$\omega \geqq \omega_0: \quad e^{2T} = \frac{\sigma+1}{\sigma-1} = \frac{\sqrt{(1-\omega_0^2/\omega^2)}+1}{\sqrt{(1-\omega_0^2/\omega^2)}-1}$$

$$= -\frac{(\omega/\omega_0)+\sqrt{(\omega^2/\omega_0^2-1)}}{(\omega/\omega_0)-\sqrt{(\omega^2/\omega_0^2-1)}}$$

Hence $T = \alpha + j\pi/2$, where $\cosh\alpha = \omega/\omega_0$. The value of $\beta$ is chosen to give continuity at $\omega = \omega_0$. The graphical form of these expressions is shown in Fig. 3(*a*) and 3(*b*).

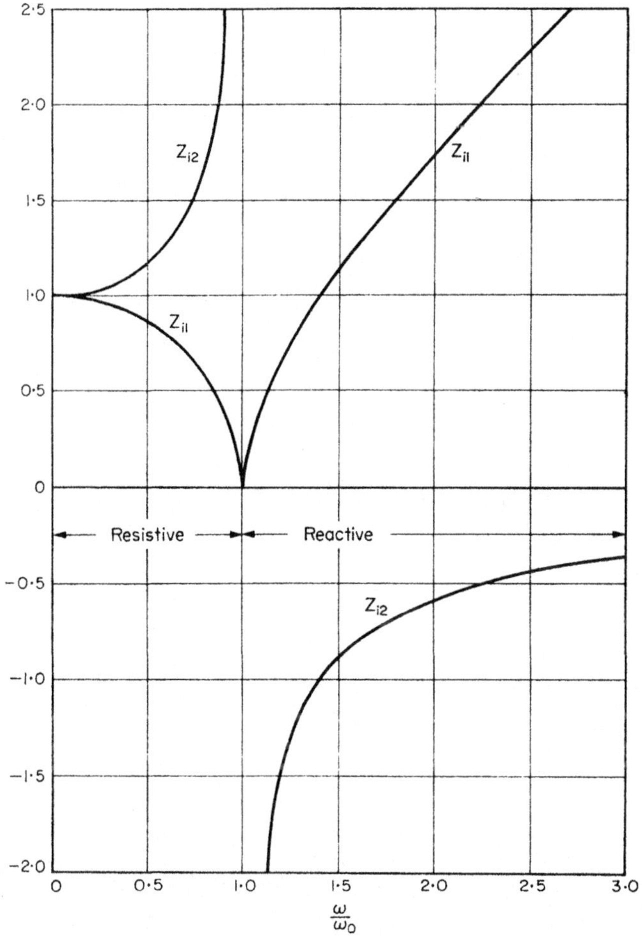


FIG. 3(*b*). *Image impedances (units of* $\sqrt{(L/C)}$*)*

## THE SIMPLE THEORY OF FILTERS

Consider the circuit of Fig. 4.

FIG. 4. *Image-terminated reactive network*

Equations (34) of Chapter 5 give in this case

$$V_1 = \lambda\sqrt{Z_{i1}}$$
$$V_2 = \lambda\sqrt{Z_{i2}}\,e^{-T}$$

where $T$ is the image propagation constant of the 2-port.

Further $\qquad V_1 = \frac{1}{2}E$

Hence $\qquad V_2 = \frac{1}{2}E\sqrt{(Z_{i2}/Z_{i1})}\,e^{-T}$

If we consider a symmetrical section ($Z_{i1} = Z_{i2}$) this reduces to

$$V_2 = \tfrac{1}{2}E\,e^{-T}$$

Hence in that range of frequencies for which $T$ is purely imaginary $|V_2| = \frac{1}{2}E$. For other frequencies $|V_2| < \frac{1}{2}E$, since $|e^{-T}| < 1$. Hence we may speak of a pass band and a stop band. In practice it is impossible exactly to realise image impedance terminations, but they may be approximately realised over

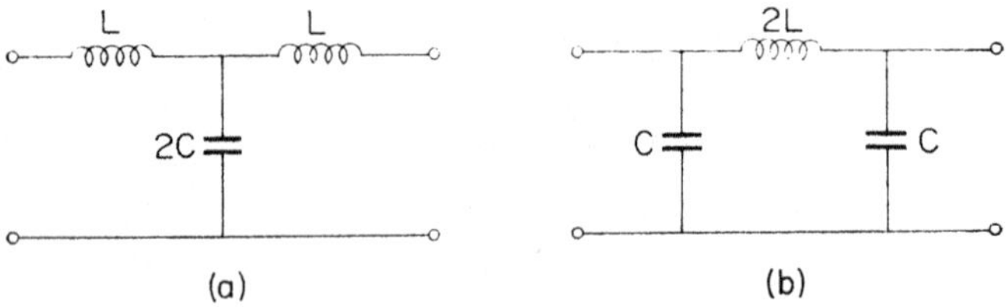


FIG. 5. *Symmetrical networks*

part of the pass band. In this range one would expect roughly constant output. At frequencies for which the output is small the terminations cannot affect the situation greatly, but at other frequencies this simple theory cannot be applied without further consideration.

*Example.* Low pass filter.

Two symmetrical circuits may be constructed from that of Fig. 2. They are shown in Fig. 5(*a*) and 5(*b*).

The image propagation constant of both will be the same, twice that of a single section. The image impedance of the T-circuit will be $Z_{i1}$ and of the Π-circuit $Z_{i2}$. Matching near zero frequency may be obtained in both cases by using a resistance $\sqrt{(L/C)}$. The nominal pass band extends from zero to $1/\sqrt{(LC)}$.

## REACTIVE 2-PORT BETWEEN RESISTIVE IMPEDANCES

When a filter is to be inserted between a source and a load, the impedances of source and load obviously have a critical effect on performance. For most practical purposes, both the source and the load may be considered to be purely resistive. There will often be reactances associated with the resistances, but for theoretical purposes it is more convenient to consider purely resistive source and load impedances and to incorporate the reactances as part of the filter circuit. This may be simply possible if the reactances are only strays, or it may greatly influence choice of filter circuit.

Consider the circuit of Fig. 6.

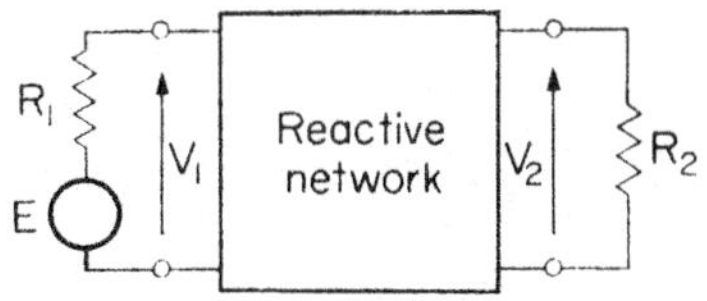


FIG. 6. *Reactive network between resistive terminations*

Without actual calculation, certain limits may be placed on the output voltage $V_2$ by consideration of power. The power into the load is

$$\frac{1}{R_2}|V_2|^2$$

This must also be the power delivered by the source to the network, since the latter cannot absorb power. Available power from the source is

$$\frac{1}{4R_1}E^2$$

which is obtained when the source is loaded with an impedance $R_1$. Hence

$$\frac{1}{R_2}|V_2|^2 \leqq \frac{E^2}{4R_1} \tag{8}$$

The equality obtains only when the terminated network presents an impedance $R_1$ to the source. One case when this is so occurs when, at a particular frequency, $R_1$ and $R_2$ are the values of the appropriate image impedances of the network. At this frequency $V_2$ would attain its maximum value. If this were the only way in which maximum power could be obtained, the chances of passing a range of frequencies would be small. It will be shown in the next section that frequencies other than those giving image matching also allow maximum power transfer.

When a source of e.m.f. $E$ and resistive impedance $R_1$ feeds a load of impedance $Z$, the power delivered to the load is

$$\left|\frac{E}{R_1+Z}\right|^2 \mathrm{Re}(Z) = \frac{E^2}{(R_1+Z)(R_1+Z^*)} \times \tfrac{1}{2}(Z+Z^*)$$

It may readily be ascertained that this expression is identical to

$$\frac{E^2}{4R_1}(1-|K|^2) \tag{9}$$

where $$K = \frac{Z-R_1}{Z+R_1} \tag{10}$$

This expression is convenient since the coefficient $E^2/4R_1$ is the available power, and $K$ is a measure of mismatch of the load impedance with the source impedance. By analogy with transmission lines $K$ is sometimes called a reflection coefficient.

## THE INPUT IMPEDANCE OF A REACTIVE NETWORK TERMINATED BY A RESISTANCE

If we consider the circuit of Fig. 6, equations (34) of Chapter 5 give the relations

$$V_1 = \sqrt{Z_{i1}}\,(\lambda+\mu)$$
$$I_1 = \sqrt{(1/Z_{i1})}(\lambda-\mu)$$
$$V_2 = \sqrt{Z_{i2}}\,(\lambda \mathrm{e}^{-T}+\mu \mathrm{e}^{T})$$
$$I_2 = \sqrt{(1/Z_{i2})}(\lambda \mathrm{e}^{-T}-\mu \mathrm{e}^{T})$$

Further, in this circuit

$$V_2/I_2 = R_2$$

Hence $$\frac{R_2}{Z_{i2}} = \frac{\lambda \mathrm{e}^{-2T}+\mu}{\lambda \mathrm{e}^{-2T}-\mu}$$

and $$\mu = \lambda \mathrm{e}^{-2T} k_2 \tag{11}$$

where $$k_2 = \frac{R_2-Z_{i2}}{R_2+Z_{i2}} \tag{12}$$

Hence $$\frac{V_1}{I_1} = Z_{i1}\frac{\lambda+\mu}{\lambda-\mu}$$

$$= Z_{i1}\frac{1+k_2 \mathrm{e}^{-2T}}{1-k_2 \mathrm{e}^{-2T}} \tag{13}$$

Since reactive networks are being considered we have two cases to consider (*a*) $Z_{i1}$ real, $k_2$ real, $T$ imaginary

(*b*) $Z_{i1}$ imaginary, $k_2$ complex and of unity modulus, $T$ with constant imaginary part and variable real part.

*Case* (*a*)

$$\text{Input impedance} = Z_a = \frac{V_1}{I_1} = Z_{i1}\frac{1+k_2\,e^{-2j\beta}}{1-k_2\,e^{-2j\beta}} \tag{14}$$

If matching with the source impedance is to be possible, $Z_a$ must be purely resistive. This is only possible if $k_2\,e^{-2j\beta}$ is a real number. Hence $2\beta$ must be an integral multiple of $\pi$.

Two cases arise depending on whether the multiple is even or odd.

(i) $$2\beta = 2n\pi, \qquad n \text{ any integer}$$
$$\text{and } e^{-2j\beta} = 1$$

Substituting in equation (14)

$$Z_a = Z_{i1}\frac{1+k_2}{1-k_2} = Z_{i1}\frac{R_2}{Z_{i2}}$$

(ii) $$2\beta = (2n-1)\pi, \qquad n \text{ any integer}$$
$$e^{-2j\beta} = -1$$
$$Z_a = \frac{Z_{i1}Z_{i2}}{R_2}$$

Hence a purely resistive impedance is presented to the source at frequencies for which the phase of the image propagation constant is either an integral multiple of $\pi$, when the impedance is $R_2 Z_{i1}/Zi_2$, or an integral multiple of $\pi$ less $\pi/2$, when the impedance is $Z_{i1}Z_{i2}/R_2$. The condition on $\beta$ of course defines a series of individual frequencies, and the values of $Z_{i1}$ and $Z_{i2}$ appropriate to these frequencies must be used.

We see that in two simple cases at least it is possible to have the same resistive impedance at a number of frequencies in the pass band: for a symmetrical circuit $Z_{i1} = Z_{i2}$, and in the first case we always have $R_2$; for what may be termed a reciprocally symmetrical circuit $Z_{i1}Z_{i2} = \text{constant}$, when the impedance will be $\text{const}/R_2$.

*Case* (*b*)

$$Z_a = jX_{i1}\frac{1+k_2\,e^{-2T}}{1-k_2\,e^{-2T}}$$

If $Z_a$ is to be resistive $(1+k_2e^{-2T})/(1-k_2\,e^{-2T})$ must be imaginary. It is easily verified that this can only be so if $|k_2\,e^{-2T}|=1$. Since $|k_2|=1$, this can only be the case if $\mathrm{Re}(T)=0$, i.e. at the junction between this frequency range and that considered in case (*a*).

We see that by suitable choice of terminating and source impedances the available power can be drawn from the source at a number of frequencies in the pass band (case (*a*)). The general picture of the response will be of the type shown in Fig. 7.

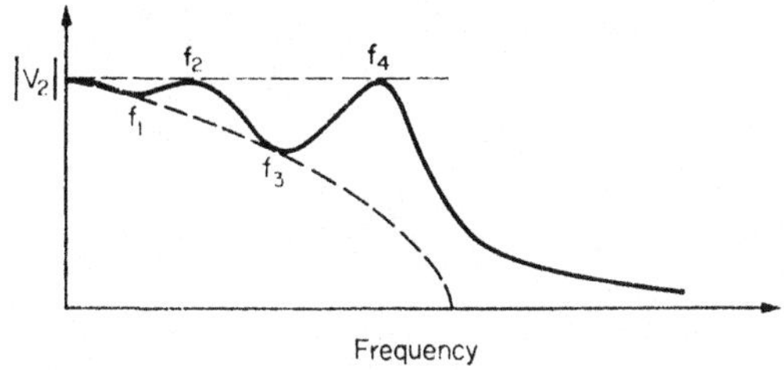


FIG. 7. *Response curve of network*

At a number of frequencies $f_2, f_4, \ldots$ maximum output will be obtained. In between minima will occur at frequencies $f_1, f_3, \ldots$. Outside the pass band the maximum cannot be attained, although we have not investigated the actual maxima that may occur.

An exact expression for the power delivered to the load will now be obtained. This power is given by equation (9) with $Z=Z_a$. Substituting in equation (10) from equation (13) we find

$$K=\frac{Z_{i1}(1+k_2\,e^{-2T})-R_1(1-k_2\,e^{-2T})}{Z_{i1}(1+k_2\,e^{-2T})+R_1(1-k_2\,e^{-2T})}$$

$$=\frac{-k_1+k_2\,e^{-2T}}{1-k_1\,k_2\,e^{-2T}} \tag{15}$$

where
$$k_1 = \frac{R_1 - Z_{i1}}{R_1 + Z_{i1}} \tag{16}$$

Algebraically simple results can only be obtained in the two special cases mentioned above: $Z_{i1} = Z_{i2}$ or $Z_{i1} Z_{i2} = R_0^2$, constant.

## SYMMETRICAL NETWORKS WITH SYMMETRICAL TERMINATION

In this case $k_1 = k_2$ and we find

$$K = -k \frac{1 - e^{-2T}}{1 - k^2 e^{-2T}}$$

Hence
$$|K|^2 = |k|^2 \left| \frac{1 - e^{-2T}}{1 - k^2 e^{-2T}} \right|^2 \tag{17}$$

In the pass band this becomes

$$|K|^2 = k^2 \left| \frac{1 - e^{-2j\beta}}{1 - k^2 e^{-2j\beta}} \right|^2$$

$$= k^2 \frac{2(1 - \cos 2\beta)}{1 - 2k^2 \cos 2\beta + k^4}$$

Hence
$$1 - |K|^2 = \frac{(1 - k^2)^2}{1 - 2k^2 \cos 2\beta + k^4} \tag{18}$$

If $\beta$ is considered to be varying with $k$ held constant, this expression lies between limits

$$\left( \frac{1 - k^2}{1 \pm k^2} \right)^2$$

Hence
$$1 \geqq 1 - |K|^2 \geqq \left( \frac{1 - k^2}{1 + k^2} \right)^2$$

The maximum is attained when $\cos 2\beta = 1$, and the minimum when $\cos 2\beta = -1$, as deduced earlier.

We may thus fill in the equation of the curve in Fig. 7 which bounds the minima. It is proportional to $(1-k^2)/(1+k^2)$. $k$ must go to unity at the cut off frequency, but at this frequency, as shown later, $e^{2j\beta}=1$. Hence equation (18) assumes a limiting form and does not correspond to a maximum of output.

In the stop band

$$|K|^2 = |k|^2 \left| \frac{1-e^{-2\alpha}}{1-k^2 e^{-2\alpha}} \right|^2 \tag{19}$$

We may write $k^2 = e^{2j\theta}$ where $\theta = \tan^{-1}|Z_{i1}/R|$ and lies between $\pm\pi/2$. In this expression we have assumed that $T=\alpha+jn\pi$: that a symmetrical circuit must have this form in the stop band has not been shown, but it will be seen to be necessarily so. (We note that the asymmetrical circuit of Fig. 2 has $T=\alpha+j\pi/2$.)

Evaluating the right-hand side of equation (19)

$$|K|^2 = \frac{(1-e^{-2\alpha})^2}{1-2e^{-2\alpha}\cos 4\theta + e^{-4\alpha}}$$

From this expression we may deduce that

$$\cosh 2\alpha = \frac{1-|K|^2 \cos 4\theta}{1-|K|^2} \tag{20}$$

For real circuits $|K| \leqq 1$, and hence for any given $K$ and $\theta$, $\alpha$ may be found. If in equation (19) we had allowed $e^{-2T} = -e^{-2\alpha}$ equation (20) would have been obtained with a negative sign in front of $\cosh 2\alpha$. Since the expression on the right-hand side is always positive and greater than unity, no real $\alpha$ would satisfy it with this modification. Alternatively one may say that with real $\alpha$ it would imply $|K| > 1$, which is impossible.

The ratio of actual power output to the available power is $1-|K|^2$. From equation (20) this is given by

$$1-|K|^2 = \frac{1-\cos 4\theta}{\cosh 2\alpha - \cos 4\theta}$$

This expression is bounded as a function of $\theta$ by zero and

$$\frac{2}{\cosh 2\alpha + 1} = \operatorname{sech}^2 \alpha \tag{21}$$

For moderately large values of $\alpha$ we may replace this by $4e^{-2\alpha}$.

Hence the maximum power output is always less than the expression (21) multiplied by the available output power.

In this section we have obtained a reasonable idea of the actual response of a symmetrical network between equal resistive loads.

The power output in the pass band of the simple theory consists of a series of ripples rising to the maximum possible output $P_0$, with minima governed by the function $P_0(1-k^2)^2/(1+k^2)^2$. In the stop band the power output is always less than $P_0 \operatorname{sech}^2 \alpha$.

The other case mentioned, $Z_{i1}Z_{i2} = R_0^2$, may be similarly treated. Terminations of $R_1$ and $R_2 = R_0^2/R_1$ give $k_1 = -k_2$, and the foregoing treatment is only slightly modified.

## *Examples on Chapter 6*

1. Obtain expressions for the image parameters for the network below.

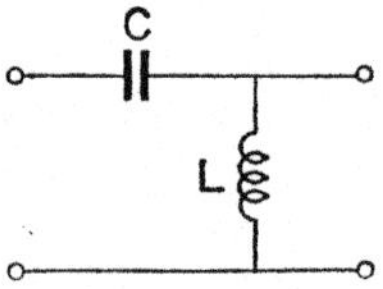


*Answer:*

$$\omega^2 LC < 1: \quad Z_{i1} = \frac{\sqrt{(1-\omega^2 LC)}}{j\omega C}, \quad Z_{i2} = \frac{j\omega L}{\sqrt{(1-\omega^2 LC)}}$$

$$e^{2T} = [\sqrt{(1-\omega^2 LC)}+1]/[\sqrt{(1-\omega^2 LC)}-1]$$

$$\omega^2 LC > 1: \quad Z_{i1} = \frac{\sqrt{(\omega^2 LC-1)}}{\omega C}, \quad Z_{i2} = \frac{\omega L}{\sqrt{(\omega^2 LC-1)}}$$

$$e^{2T} = [j\sqrt{(\omega^2 LC-1)}+1]/[j\sqrt{(\omega^2 LC-1)}-1]$$

2. For the network below
   (*a*) find the frequencies for which maximum transmission takes place
   (*b*) the output power for $\omega\sqrt{(LC)} = 1$
   (*c*) the function $[(1-k^2)/(1+k^2)]^2$
   (*d*) the function $P_0 \operatorname{sech}^2 \alpha$.

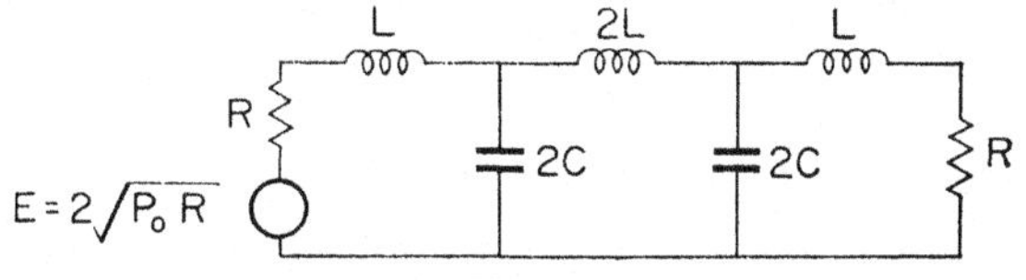


$R = \sqrt{\frac{L}{C}}$

*Answer:* If $x = \omega\sqrt{(LC)}$,

(*a*) $x = 0,\ 1/\sqrt{2}$
(*b*) $P_0/5$
(*c*) $(1-x^2)/(1-x^2+\frac{1}{4}x^4)$
(*d*) $P_0/(8x^4-8x^2+1)^2$

3. A reactive network of the type considered in this chapter is terminated in a resistance $R$ and fed from a source of e.m.f. $E$ and of zero impedance. Show that in the pass band of the network the power supplied to the load is

$$(E^2/Z_{i1})(1-k_2^2)/(1+2k_2 \cos 2\beta + k_2^2)$$

Find the values this takes for $\cos 2\beta = \pm 1$.

*Answer:*

$$(E^2/R)(Z_{i2}/Z_{i1}); \quad E^2R/Z_{i1} Z_{i2}$$

# 7 FILTER CIRCUITS

IN the previous chapter the theory of a reactive 2-port network operating between resistive impedances has been given. It is the purpose of the present chapter to give some of the circuits which may be used.

## LADDER NETWORKS

Ladder networks are most conveniently analysed in terms of an L-section, as shown in Fig. 1(*a*). By placing sections back-to-back or front-to-front, the T- and Π-circuits of Fig. 1(*b*) and 1(*c*) are

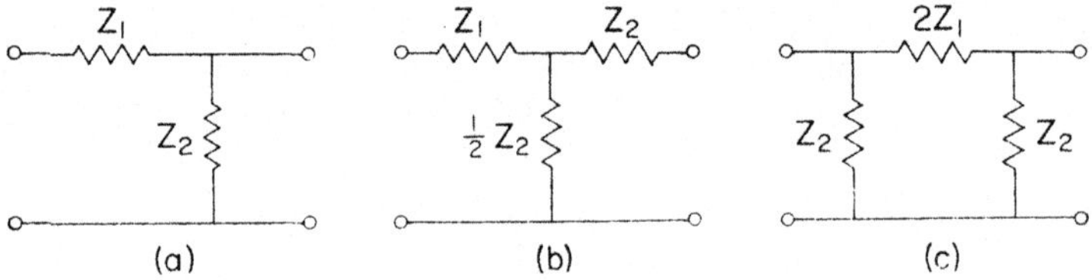


FIG. 1. *Elementary filter circuit*

obtained. The image impedances of these will be those of the corresponding ends of the L-circuit and, since the network is reciprocal, the image propagation constant will be twice that of the L-circuit.

The open- and short-circuit impedances for the circuit of Fig. 1(*a*) are

$$Z_{o1} = Z_1 + Z_2$$
$$Z_{s1} = Z_1$$
$$Z_{o2} = Z_2$$
$$Z_{s2} = Z_1 Z_2/(Z_1 + Z_2)$$

Hence $$Z_o/Z_s = 1 + Z_2/Z_1 \qquad (1)$$

We then have

$$\left.\begin{aligned} Z_{i1} &= Z_1 \sqrt{(1+Z_2/Z_1)} \\ Z_{i2} &= Z_1 Z_2 \sqrt{(1+Z_2/Z_1)}/(Z_1+Z_2) \\ &= Z_2/\sqrt{(1+Z_2/Z_1)} \end{aligned}\right\} \qquad (2)$$

$$e^{2T} = \frac{\sqrt{(1+Z_2/Z_1)}+1}{\sqrt{(1+Z_2/Z_1)}-1} \qquad (3)$$

In these expressions the sign of the radical has to be chosen as discussed earlier. Useful variants of equation (3) are

$$\cosh 2T = 1 + 2Z_1/Z_2 \qquad (4)$$

$$\tanh T = 1/\sqrt{(1+Z_2/Z_1)} \qquad (5)$$

A particular class of this circuit is that for which $Z_1$ and $Z_2$ satisfy the relation

$$Z_1 Z_2 = R_0^2 \qquad (6)$$

in which $R_0$ is independent of frequency. The case discussed in Chapter 6 with $Z_1$, $Z_2$, pure inductance and pure capacitance respectively, comes in this category. The formulae then become slightly simpler in that only one impedance enters into the expressions.

As an example more complicated than the simple *L–C* case consider the circuit of Fig. 2.

Here $$Z_1 = j(\omega L_1 - 1/\omega C_1)$$
$$1/Z_2 = j(\omega C_2 - 1/\omega L_2)$$

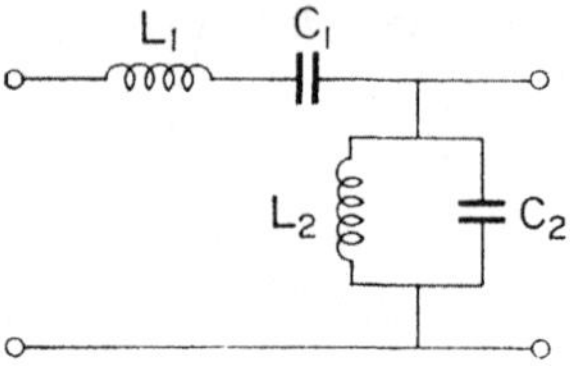


FIG. 2. *Band-pass filter*

If equation (6) is to hold we must have

$$1/\sqrt{(L_1 C_1)} = 1/\sqrt{(L_2 C_2)} = \omega_0$$

$$Z_1 = j\sqrt{\left(\frac{L_1}{C_1}\right)}\left(\frac{\omega}{\omega_0} - \frac{\omega_0}{\omega}\right) = jR_1\left(\frac{\omega}{\omega_0} - \frac{\omega_0}{\omega}\right)$$

$$Z_2 = -j\sqrt{\left(\frac{L_2}{C_2}\right)}\left(\frac{\omega}{\omega_0} - \frac{\omega_0}{\omega}\right)^{-1} = -jR_2\left(\frac{\omega}{\omega_0} - \frac{\omega_0}{\omega}\right)^{-1}$$

$$Z_1 Z_2 = R_1 R_2 = R_0^2$$

where the symbols $R_1$ and $R_2$ have been introduced for $\sqrt{(L_1/C_1)}$, $\sqrt{(L_2/C_2)}$ respectively.

From equation (1)

$$\frac{Z_o}{Z_s} = 1 - \frac{R_2}{R_1}\left(\frac{\omega}{\omega_0} - \frac{\omega_0}{\omega}\right)^{-2}$$

The cut-off frequencies $\omega_1$, $\omega_2$ occur where this function becomes zero. Hence $\omega_1$ and $\omega_2$ satisfy the equations

$$\frac{\omega}{\omega_0} - \frac{\omega_0}{\omega} = \pm\sqrt{\left(\frac{R_2}{R_1}\right)}$$

or

$$\left(\frac{\omega}{\omega_0}\right)^2 \pm 2\frac{\omega}{\omega_0}\sqrt{\left(\frac{R_2}{R_1}\right)} - 1 = 0$$

It can be seen that the four possible roots are $\pm\omega_1$, $\pm\omega_2$, occurring so that with one sign of the radical we have $\omega_1$, $-\omega_2$, and with the other $-\omega_1$, $\omega_2$. Since we are usually concerned with design rather than analysis, it is convenient to regard $\omega_1$, $\omega_2$ as given. We then have, from the properties of the roots of a quadratic equation, assuming $\omega_2 > \omega_1$,

$$\omega_0^2 = \omega_1 \omega_2$$

$$2\sqrt{\left(\frac{R_2}{R_1}\right)} = \frac{\omega_2 - \omega_1}{\omega_0}$$

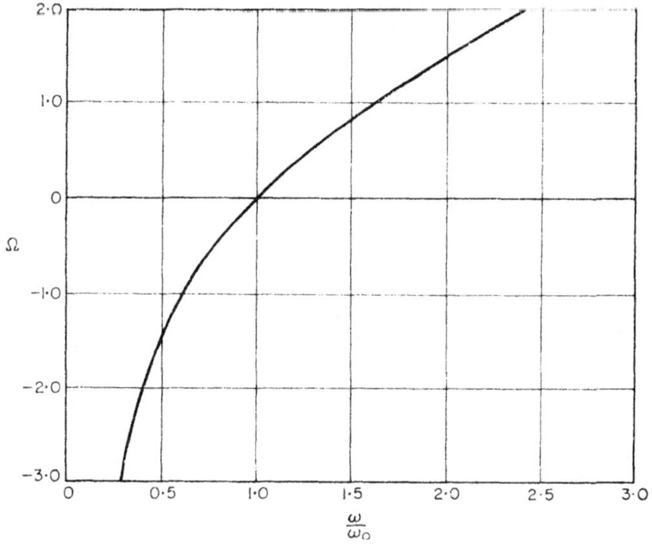


FIG. 3. *The function* $\Omega = \dfrac{\omega}{\omega_0} - \dfrac{\omega_0}{\omega}$

It may be noted that this latter equation can be written in the form

$$C_2 R_0 (\omega_2 - \omega_1) = 1$$

The expressions for image impedance become simpler if the following notation is introduced

$$\Omega = \frac{\omega}{\omega_0} - \frac{\omega_0}{\omega}$$

$$2\delta = \frac{\omega_2 - \omega_1}{\omega_0}$$

$\delta$ is the fractional half-bandwidth. A graph of $\Omega$ as a function of $\omega$ is shown in Fig. 3.

We then have

$$R_1 = R_0/\delta$$
$$R_2 = R_0\,\delta$$
$$Z_1 = jR_0\,\Omega/\delta$$
$$Z_2 = -j\,\delta R_0/\Omega$$
$$Z_2/Z_1 = -\delta^2/\Omega^2$$

The image impedances are given by the expressions

$$Z_{i1} = j\frac{R_0\,\Omega}{\delta}(1-\delta^2/\Omega^2)^{\frac{1}{2}}$$

$$Z_{i2} = -j\frac{\delta R_0}{\Omega}(1-\delta^2/\Omega^2)^{-\frac{1}{2}}$$

The pass band corresponds to $|\Omega| \leqq \delta$. For positive resistive impedances in the pass band we must choose

$$\left(1-\frac{\delta^2}{\Omega^2}\right)^{\frac{1}{2}} = -j\sqrt{\left(\frac{\delta^2}{\Omega^2}-1\right)}, \qquad \Omega > 0$$

$$\left(1-\frac{\delta^2}{\Omega^2}\right)^{\frac{1}{2}} = j\sqrt{\left(\frac{\delta^2}{\Omega^2}-1\right)}, \qquad \Omega < 0$$

In the stop band

$$\left(1-\frac{\delta^2}{\Omega^2}\right)^{\frac{1}{2}} = \sqrt{\left(1-\frac{\delta^2}{\Omega^2}\right)}, \qquad |\Omega| > \delta$$

In these expressions the radical is to be taken positive.

The form of the image impedances as functions of frequency are illustrated graphically in Fig. 4.

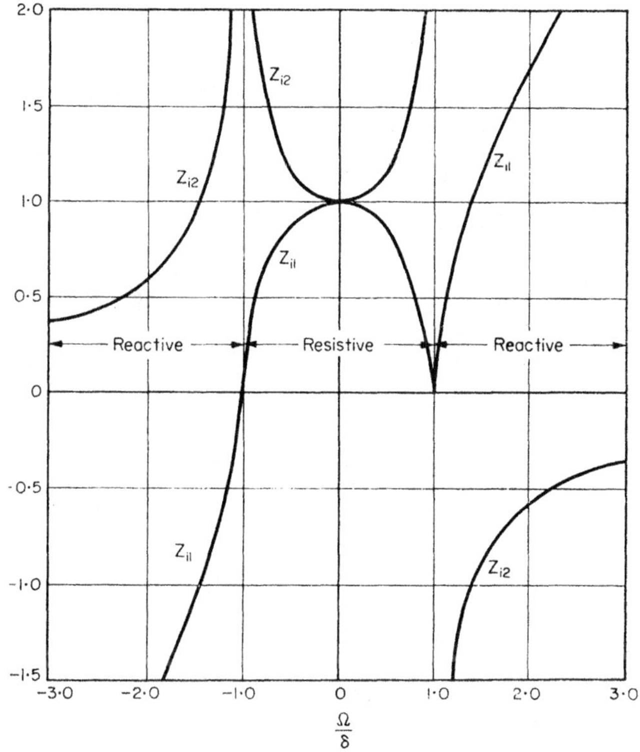


FIG. 4. *Image impedances (units of $R_0$). Note that abscissae are not frequency*

The expression for the propagation constant is

$$e^{2T} = \frac{(1-\delta^2/\Omega^2)^{\frac{1}{2}}+1}{(1-\delta^2/\Omega^2)^{\frac{1}{2}}-1}$$

In the pass band $T = j\beta$ and we have

$$e^{2j\beta} = \frac{\pm j\sqrt{(\delta^2/\Omega^2 - 1)} + 1}{\pm j\sqrt{(\delta^2/\Omega^2 - 1)} - 1} = \frac{\sqrt{(1 - \delta^2/\Omega^2)} \mp j(|\Omega|/\delta)}{\sqrt{(1 - \delta^2/\Omega^2)} \pm j(|\Omega|/\delta)}$$

in which the top sign is for $\omega < \omega_0$ $(\Omega < 0)$ and the lower for $\omega > \omega_0$ $(\Omega > 0)$. We may incorporate the two signs by deleting the modulus and hence

$$\beta = \sin^{-1}\frac{\Omega}{\delta}$$

In the stop band

$$e^{2T} = \frac{\sqrt{(1 - \delta^2/\Omega^2)} + 1}{\sqrt{(1 - \delta^2/\Omega^2)} - 1} = -\frac{|\Omega|/\delta + \sqrt{(\Omega^2/\delta^2 - 1)}}{|\Omega|/\delta - \sqrt{(\Omega^2/\delta^2 - 1)}}$$

Making the phase continuous at the cut-off frequencies we find

$$T = \cosh^{-1}\frac{|\Omega|}{\delta} \pm j\frac{\pi}{2}, \qquad \begin{array}{l} +,\ \omega > \omega_0 \\ -,\ \omega < \omega_0 \end{array}$$

$T$ is shown graphically in Fig. 5.

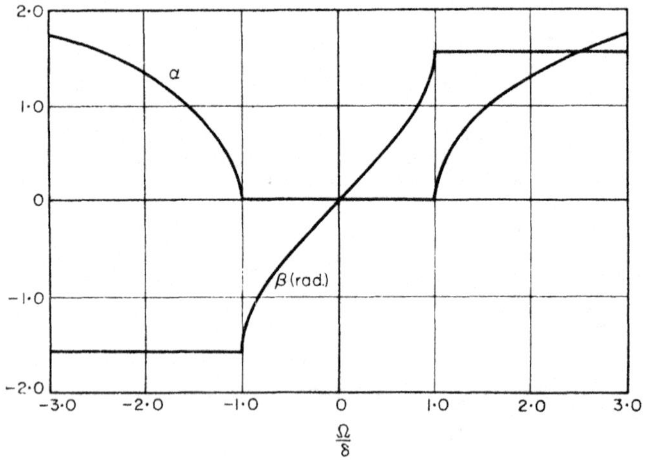


FIG. 5. *Attenuation and phase*

## $m_T$-DERIVED SECTIONS

If we wish to join different sections together we must preserve image matching at the junctions. The simple sections, such as the two examples considered, give an attenuation which increases steadily away from the pass band, but do not admit of a maximum attenuation at some finite frequency. In seeking for other sections with different properties, we may inquire if it is possible to have two L-sections which have the same image impedance on one side but different on the other.

Consider the two sections of Fig. 6(*a*) and 6(*b*).

Equal image impedances at the input will occur if

$$Z_{i1} = \sqrt{(Z_{o1} Z_{s1})} = \sqrt{(Z'_{o1} Z'_{s1})}$$

This is satisfied by

$$Z'_{s1} = mZ_{s1} \tag{7}$$

$$Z'_{o1} = (1/m)Z_{o1} \tag{8}$$

Equation (7) leads immediately to the relation

$$Z'_1 = mZ_1 \tag{9}$$

Equation (8) then gives

$$Z'_2 = \frac{Z_2}{m} + \left(\frac{1}{m} - m\right)Z_1 \tag{10}$$

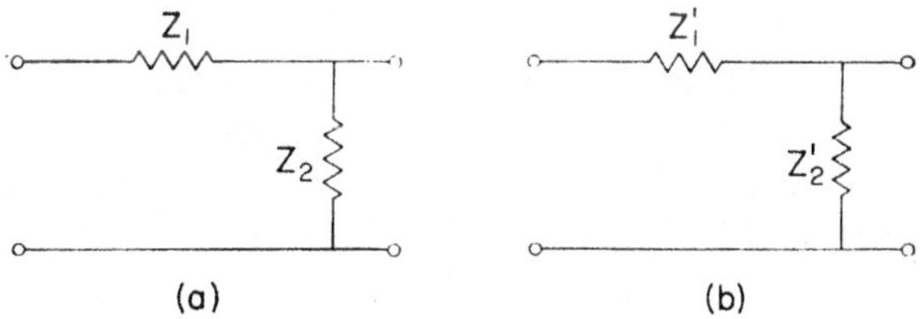


FIG. 6. *Elementary L-sections*

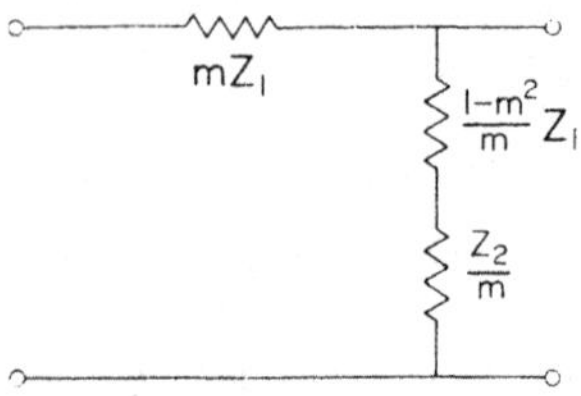


FIG. 7. *$m_T$-derived section*

Since it is not usually possible to realise a negative multiple of a physical impedance, $m$ must be restricted to a positive number less than unity. The components of $Z_1'$ and $Z_2'$ are then positive multiples of $Z_1$ and $Z_2$.

We can now proceed to investigate the other image impedance and the propagation constant. The section under investigation is that of Fig. 7, and is called the $m_T$-derived section of that of Fig. 6($a$).

We have

$$Z_{o2}' = \frac{1-m^2}{m}Z_1 + \frac{Z_2}{m}$$

$$Z_{s2}' = m\{(1-m^2)Z_1 + Z_2\}(1+Z_2/Z_1)^{-1}$$

Hence
$$\frac{Z_o'}{Z_s'} = \frac{1}{m^2}\left(1+\frac{Z_2}{Z_1}\right) \tag{11}$$

and
$$\begin{aligned} Z_{i2}' &= \{(1-m^2)Z_1 + Z_2\}(1+Z_2/Z_1)^{-\frac{1}{2}} \\ &= Z_{i2}\{1+(1-m^2)Z_1/Z_2\} \end{aligned} \tag{12}$$

We see that the ratio $Z_o'/Z_s'$ is a positive multiple of $Z_o/Z_s$ for the original section. This must be so since the equality of the image impedances ensures the same pass band.

The propagation constant is given by the expression

$$e^{2T} = \frac{(1+Z_2/Z_1)^{\frac{1}{2}}+m}{(1+Z_2/Z_1)^{\frac{1}{2}}-m} \tag{13}$$

## THE LOW-PASS FILTER

Consider the low-pass filter section considered in Chapter 6. The original section and $m_T$-derived section are shown in Figs. 8(*a*) and 8(*b*).

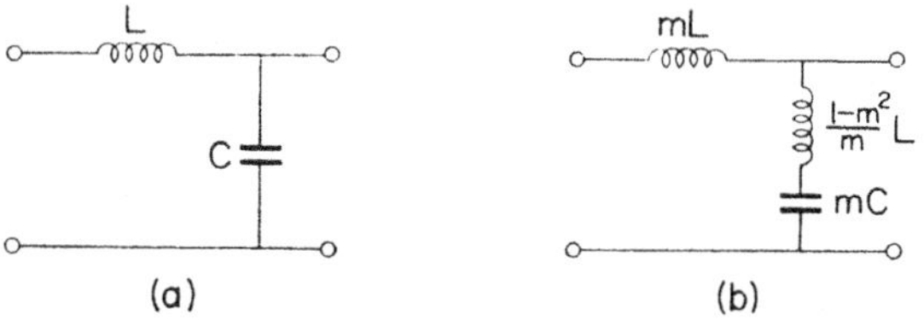


FIG. 8. *$m_T$-derived low-pass filter section*

For the original section it has been shown (Chapter 6) that

$$Z_{i2} = \sqrt{\frac{L/C}{1-\omega^2/\omega_0^2}}, \quad 0 < \omega < \omega_0$$
$$= -j\sqrt{\frac{L/C}{\omega^2/\omega_0^2-1}}, \quad \omega > \omega_0$$

For this case

$$(1-m^2)Z_1/Z_2 = -(1-m^2)\omega^2/\omega_0^2$$

Combining these results according to equation (12) we have

$$Z'_{i2} = \sqrt{\left(\frac{L}{C}\right)}\frac{1-(1-m^2)\omega^2/\omega_0^2}{\sqrt{(1-\omega^2/\omega_0^2)}}, \quad 0 < \omega < \omega_0$$
$$= -j\sqrt{\left(\frac{L}{C}\right)}\frac{1-(1-m^2)\omega^2/\omega_0^2}{\sqrt{(\omega^2/\omega_0^2-1)}}, \quad \omega > \omega_0$$

These functions are shown graphically in Fig. 9.

The propagation constant, using the results of Chapter 6, is given by the expressions

$$e^{2T} = \frac{\sqrt{(1-\omega^2/\omega_0^2)}+jm\omega/\omega_0}{\sqrt{(1-\omega^2/\omega_0^2)}-jm\omega/\omega_0}, \qquad 0 < \omega < \omega_0$$

$$= -\frac{m\omega/\omega_0+\sqrt{(\omega^2/\omega_0^2-1)}}{m\omega/\omega_0-\sqrt{(\omega^2/\omega_0^2-1)}}, \qquad \omega < \omega_0$$

Hence in the pass band $T = j\beta$ where

$$\sin\beta = (m\omega/\omega_0)\{1-(1-m^2)\omega^2/\omega_0^2\}^{-\frac{1}{2}}$$

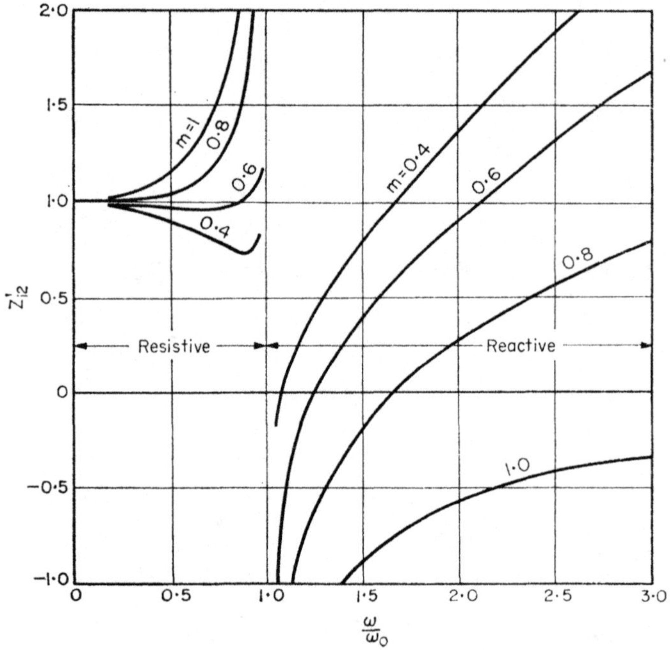


FIG. 9. *Π-image impedance for an $m_T$-derived section (units of $\sqrt{(L/C)}$)*

In the stop band a new phenomenon is exhibited: the expression for $\exp 2T$ becomes infinite at a finite frequency $\omega_1$, given by

$$\frac{\omega_1^2}{\omega_0^2} = \frac{1}{1-m^2}$$

At this frequency the attenuation is infinite, and of course it corresponds to the series resonance of the shunt arm of Fig. 8(*b*). The complete expressions for $T$ in the stop band are

$$T = \cosh^{-1}[(m\omega/\omega_0)\{1-(1-m^2)\omega^2/\omega_0^2\}^{-\frac{1}{2}}]+j\pi/2, \quad \omega_0 < \omega < \omega_1$$
$$= \sinh^{-1}[(m\omega/\omega_0)\{(1-m^2)(\omega^2/\omega_0^2)-1\}^{-\frac{1}{2}}], \quad \omega > \omega_1$$

These expressions are illustrated in Fig. 10 for $m = 0{\cdot}6$.

We see that the $m_T$-derived section has two useful properties not possessed by the original section: it gives infinite attenuation at a finite frequency outside the pass band, and it has one image

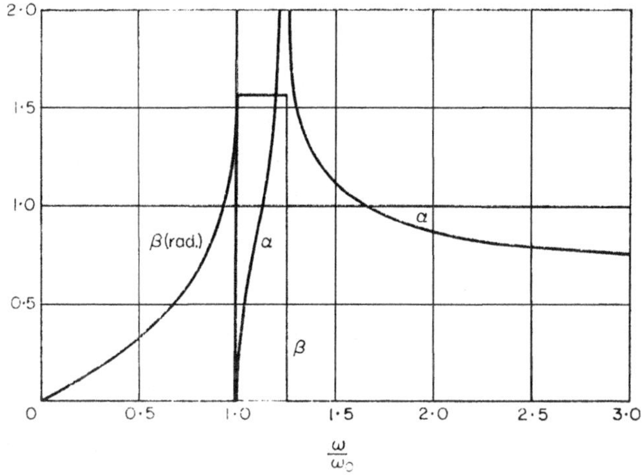


FIG. 10. *Attenuation and phase for $m_T$-derived section* ($m = 0{\cdot}6$)

impedance which may be made by suitable choice of $m$ very nearly constant over a large fraction of the pass band. A typical low-pass filter might be made up as indicated in Fig. 11.

It would begin and end with half-sections (1 and 4) chosen with $m$ about 0·6 in order to obtain matching with constant source and load resistances. It would contain one or more full T-sections of the $m$-derived type (2) in order to give high attenuation at one or more prescribed frequencies. It would also contain a number of ordinary T-sections (3) to provide adequate attenuation over the whole of the stop band.

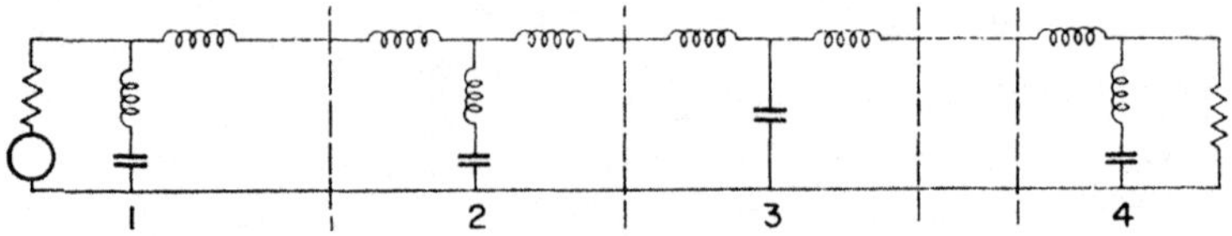


FIG. 11. *Typical low-pass filter*

## $m_\pi$-DERIVED SECTIONS

The whole work of the previous section can be repeated with $Z_{i2}$ of the original section and the new section kept the same. This leads to

$$Z'_{o2} = Z_{o2}/m$$
$$Z'_{s2} = mZ_{s2}$$

Hence
$$Z'_2 = Z_2/m \tag{14}$$

and
$$\frac{1}{Z'_1} = \frac{1}{m}\left(\frac{1}{Z_1}+\frac{1}{Z_2}\right)-\frac{m}{Z_2}$$
$$=\frac{1}{mZ_1}+\frac{1-m^2}{mZ_2} \tag{15}$$

A new section is thus developed as shown in Fig. 12.

The properties of this section are similar to those of the $m_T$-derived section. The propagation constant is identical since the ratio $Z'_o/Z'_s$ is identical in the two cases. The other image impedance is given by the expression

$$Z'_{i1} = Z_{i1}\{1+(1-m^2)Z_1/Z_2\}^{-\frac{1}{2}} \tag{16}$$

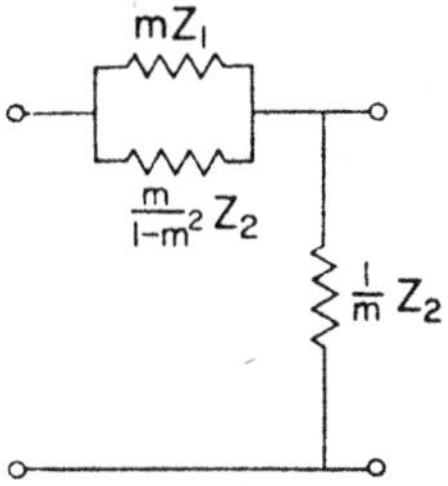


FIG. 12. *$m_\pi$-derived section*

## LATTICE FILTERS

It is a requirement in a ladder network giving high attenuation at extreme frequencies that the elements shall behave differently, as for example in the low-pass filter when one element is inductive and one capacitative. If they are both the same type, as for example in the $m_T$-derived section of Fig. 8(*b*) the attenuation at high frequencies is constant, and does not tend to infinity. The inherent losses in inductance–capacitor filters makes it impossible to obtain small fractional bandwidths, and low-loss elements are

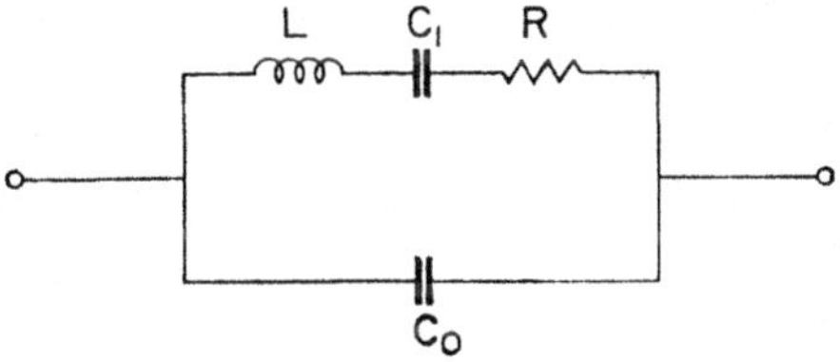


FIG. 13. *Equivalent circuit for a piezo-electric crystal*

required. Such an element is the piezo-electric crystal, which utilises the fact that certain crystals when stressed mechanically produce an electric charge separation. These crystals may be made resonant by choice of mechanical parameters, and the losses are very small in comparison with electrical elements. The $Q$-factor of a quartz crystal is quite customarily of the order of $10^5$, and by suitable mounting this may be made ten times greater. The equivalent circuit of a quartz crystal near a single resonant frequency is shown in Fig. 13.

The $L$–$C_1$–$R$ series circuit is the electrical equivalent of the mechanical motion. The shunt capacitor $C_0$ is interelectrode capacity, and cannot be eliminated. In fact, for a given type of

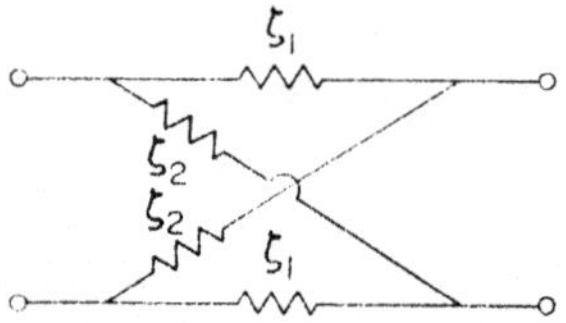


FIG. 14. *Balanced lattice section*

crystal, the ratio $C_0/C_1$ has a definite lower limit. This type of element will always look capacitative at high or low frequencies, and hence a ladder network is not suitable for a filter designed around such an element.

A suitable alternative is the lattice filter, which is almost invariably treated in its balanced form as shown in Fig. 14. The advantage of this circuit is that it obtains attenuation by a balancing of element values rather than by those values becoming very large or very small.

Here we have

$$\left.\begin{aligned} Z_{o1} &= Z_{o2} = \tfrac{1}{2}(\zeta_1+\zeta_2) \\ Z_{s1} &= Z_{s2} = 2\frac{\zeta_1\zeta_2}{\zeta_1+\zeta_2} \end{aligned}\right\} \qquad (17)$$

Hence
$$\frac{Z_o}{Z_s} = \frac{1}{4}\left(\frac{\zeta_1+\zeta_2}{\zeta_1\zeta_2}\right)^2$$

The propagation constant is given by the expression

$$e^{2T} = \frac{\frac{1}{2}(\zeta_1+\zeta_2)(\zeta_1\zeta_2)^{-\frac{1}{2}}+1}{\frac{1}{2}(\zeta_1+\zeta_2)(\zeta_1\zeta_2)^{-\frac{1}{2}}-1} \tag{18}$$

A very compact expression may be obtained for $\tanh \frac{1}{2}T$.

$$\tanh T = \frac{e^{2T}-1}{e^{2T}+1} = \frac{(\zeta_1\zeta_2)^{\frac{1}{2}}}{\zeta_1+\zeta_2} \tag{19}$$

This is satisfied by

$$\left.\begin{aligned} &\tanh \tfrac{1}{2}T = (\zeta_1/\zeta_2)^{\frac{1}{2}} \\ \text{or} \quad &\tanh \tfrac{1}{2}T = (\zeta_2/\zeta_1)^{\frac{1}{2}} \end{aligned}\right\} \tag{20}$$

The difference between these two forms is merely one of $j\pi$ in the $T$'s. Since we started from $\exp(2T)$, this has not been taken into account, and must be dealt with as in previous examples.

The pass band of this structure occurs for those frequencies for which $\zeta_1$ and $\zeta_2$ (both purely reactive) are of opposite sign. When this is so $(\zeta_1\zeta_2)^{\frac{1}{2}}$ is a real number and equations (18) and (19)

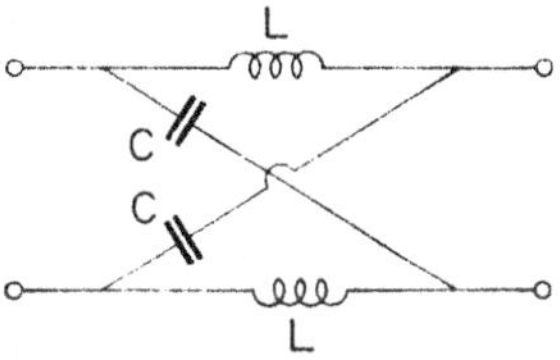


FIG. 15. *All-pass lattice section*

show that the real part of $T$ is zero. When they are of the same sign $(\zeta_1\zeta_2)^{\frac{1}{2}}$ is imaginary, and $T$ has a real part.

*Example.* Consider the circuit of Fig. 15.

Here $\zeta_1$ and $\zeta_2$ are always of opposite sign, and hence the pass band covers the whole frequency range.

From equations (17)

$$Z_i = \sqrt{(L/C)}$$

This is independent of frequency, and the image impedance is realisable as a constant resistance.

Equation (20) gives, using the same value of the radical

$$\tanh \tfrac{1}{2}T = j\omega\sqrt{(LC)}$$

Taking the phase as zero at $\omega = 0$, we see that $T = j\beta$ where

$$\beta = 2\tan^{-1}\omega\sqrt{(LC)}$$

It may be noted that this circuit, having a realisable image impedance, can be correctly terminated in practice, and when this is done it gives a phase change with no amplitude change at the output with a constant amplitude input varying in frequency. This could not happen with a ladder network.

## LOW-PASS FILTER

The design of a lattice filter to have prescribed pass bands is very simple in principle: it is only necessary to select reactances which have the required sign pattern. If we make $\zeta_2$ a capacity

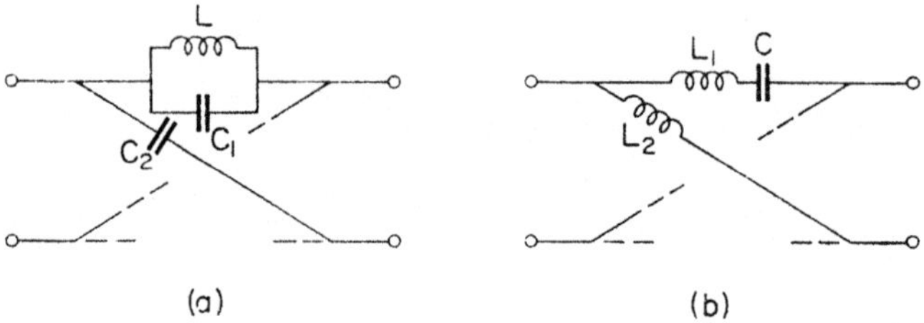


FIG. 16. *Low-pass filter section*

as in Fig. 15, and in place of $L$ use a parallel $L$–$C$ circuit, we shall have a low-pass filter. The reactances will be opposite in sign up to the resonant frequency of the parallel arm and the same sign for higher frequencies. This circuit is shown in Fig. 16($a$), and an alternative in Fig. 16($b$).

Band-pass structures are shown in Fig. 17(*a*) and 17(*b*). The pass bands of these circuits lie between the resonant frequencies of the two resonant circuits.

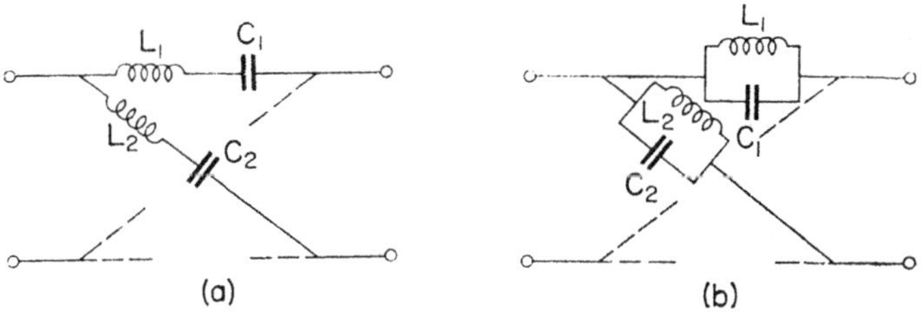


FIG. 17. *Band-pass filter section*

## THE QUARTZ CRYSTAL FILTER

A simple quartz crystal filter is shown in Fig. 18, together with its equivalent circuit.

The behaviour of the reactance of the two arms is shown graphically in Fig. 19.

The quartz crystal is characterised by a zero at $\omega_1 = 1/\sqrt{(L_1 C_1)}$ and an infinity at $\omega_2$ given by

$$\omega_2 = \sqrt{\frac{1}{L}\left[\left(\frac{1}{C_0}+\frac{1}{C_1}\right)\right]}$$

The ratio $\omega_2/\omega_1$ depends only on the capacities and is given by

$$\frac{\omega_2}{\omega_1} = \sqrt{\left(1+\frac{C_1}{C_0}\right)}$$

From this we see that to obtain the maximum difference between the two cut-off frequencies, $C_1/C_0$ must be as large as possible. This ratio has a maximum value for a given cut of crystal and is typically less than 1/125.

For low frequencies the reactance approaches that of a capacity of $C_1$ and $C_0$ in parallel, at high frequencies a capacity $C_0$. Peaks of attenuation will occur when the reactances in the two arms are equal, i.e. the bridge is balanced. With this circuit at most one peak can occur. If $C_2 < C_0$ this will be above the pass band, if $C_2 > C_1 C_0/(C_1 + C_0)$ it will occur on the lower side.

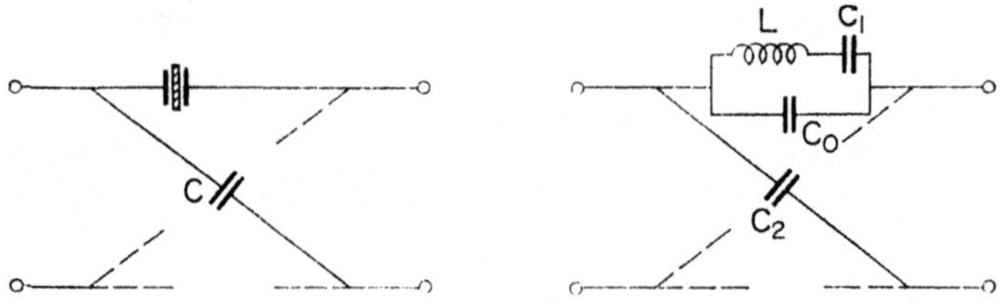


FIG. 18. *Filter section using two piezo-electric crystals*

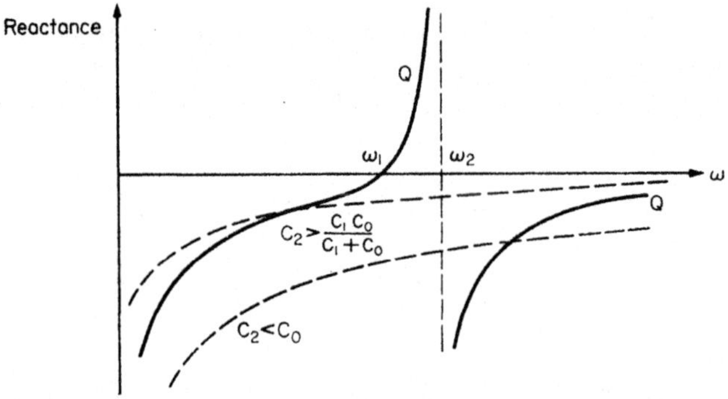


FIG. 19. *Reactance of crystal and condenser (schematic only)*

A band-pass circuit giving two peaks of attenuation is shown in Fig. 20.

The reactance pattern for a single pass band must be as shown in Fig. 21, with the zeros of one pair of crystal coinciding with the infinities of the other pair.

The pass band is between $\omega_1$ and $\omega_3$. The necessary condition for an attenuation peak on the upper side is that the $C_0$ of $Q_2$ shall be greater than the $C_0$ of $Q_1$. This will also ensure that an intersection occurs on the lower side. Equal values of $C_0$ will give infinite attenuation at zero and infinite frequency.

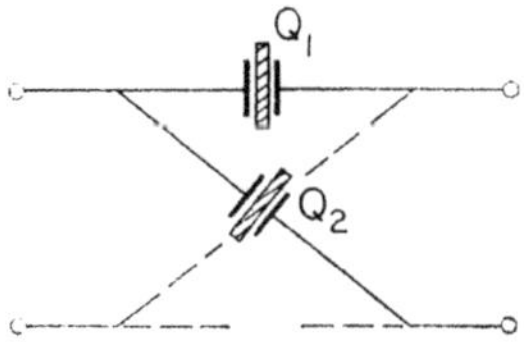


FIG. 20. *Band-pass section using four crystals*

Steeper cut-offs may be obtained by cascading several sections. A practical difficulty with these lattice filters is that they are balanced. The lack of common input and output terminals is a serious drawback at high frequencies. At frequencies in the tens

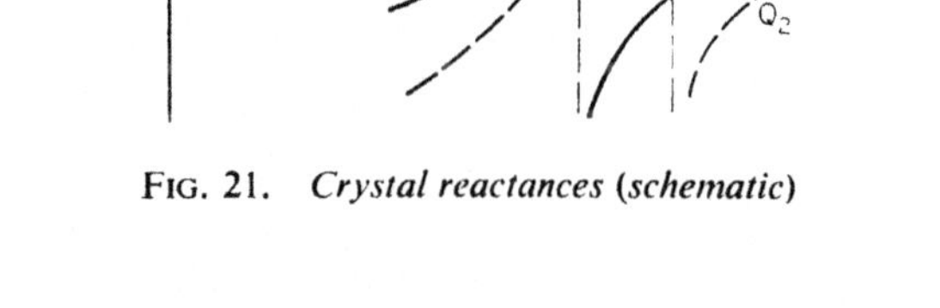


FIG. 21. *Crystal reactances (schematic)*

of megacycles per second region, circuits of the type of Fig. 22 are used. This circuit may be shown to be equivalent to two identical lattices of the type of Fig. 20 in cascade. An adequate approximation to the ideal transformer can be realised.

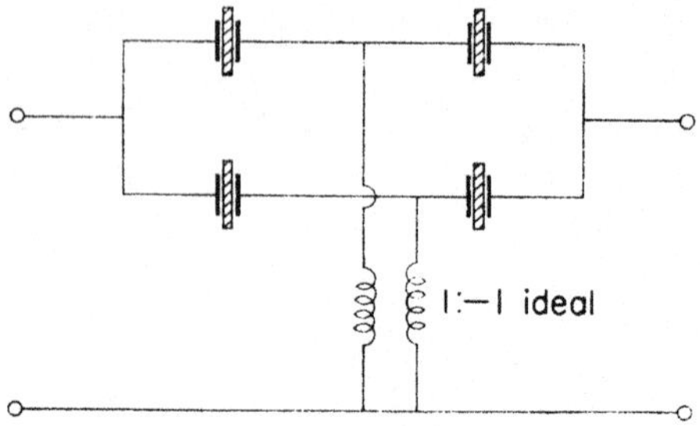


FIG. 22. *Unbalanced crystal filter*

## SINGLE SECTION FILTERS

It may be briefly pointed out that the theory can be usefully applied to single section filters such as the coupling circuits of intermediate frequency amplifiers. A typical example is that shown in Fig. 23.

This is the familiar tuned transformer circuit which may be analysed by straightforward methods. These methods yield simple results only for high-$Q$ coils, and narrow fractional bandwidths. For wide bands the algebra is too cumbersome to be useful.

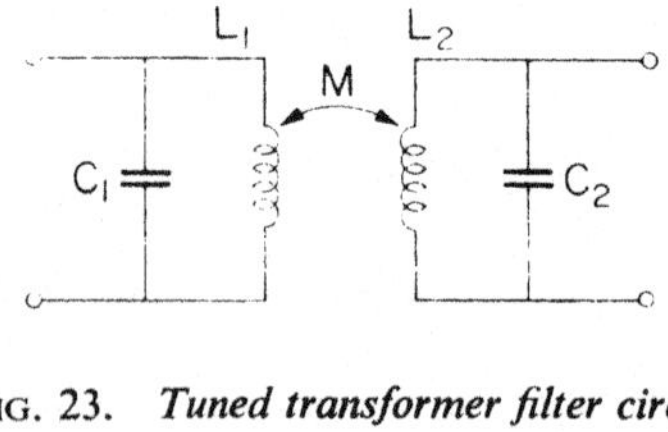


FIG. 23. *Tuned transformer filter circuit*

If losses in the coils are regarded as shunt resistors which will ultimately be part of the terminating impedances, we may consider the circuit of Fig. 23 to be reactive and analyse it as a filter structure.

We have

$$\frac{1}{Z_{o1}} = j\omega C_1 + \left[ j\omega L_1 + \frac{\omega^2 M^2}{j\omega L_2 + 1/j\omega C_2} \right]^{-1} \tag{21}$$

$$\frac{1}{Z_{s1}} = j\omega C_1 + \left[ j\omega L_1 - j\frac{\omega M^2}{L_2} \right]^{-1} \tag{22}$$

If we write

$$k^2 = M^2/L_1 L_2, \quad k_1^2 = 1 - k^2$$

$$\omega_1^2 = 1/L_1 C_1, \qquad \omega_2^2 = 1/L_2 C_2$$

we find that equations (21) and (22) may be written

$$\frac{1}{Z_{o1}} = \frac{jC_1}{\omega} \frac{\omega^4 k_1^2 - \omega^2(\omega_1^2 + \omega_2^2) + \omega_1^2 \omega_2^2}{\omega^2 k_1^2 - \omega_2^2} \tag{23}$$

$$\frac{1}{Z_{s1}} = \frac{jC_1}{\omega k_1^2}(\omega^2 k_1^2 - \omega_1^2) \tag{24}$$

Hence $$\frac{Z_s}{Z_o} = k_1^2 \frac{\omega^4 k_1^2 - \omega^2(\omega_1^2 + \omega_2^2) + \omega_1^2 \omega_2^2}{(\omega^2 k_1^2 - \omega_1^2)(\omega^2 k_1^2 - \omega_2^2)} \tag{25}$$

Equation (25) in general indicates two separate pass bands, depending on the roots of the quadratic. If we desire one pass band only, one of two things must happen: either the numerator or the denominator must be a perfect square. It can readily be shown by putting $\omega = \omega_1/k_1, \omega_2/k_1$ in the numerator that the roots of that expression lie outside the range $\omega_1/k_1 < \omega < \omega_2/k_1$, showing that we must have the denominator a perfect square.

Hence $\omega_1 = \omega_2 = \omega_0$, and the cut-off frequencies are given by

$$\omega^4 k_1^2 - 2\omega^2\omega_0^2 - \omega_0^4 = 0 \tag{26}$$

If we designate these frequencies as $\omega_a$, $\omega_b$ we have

$$\omega_a\omega_b = \omega_0^2/k_1 \tag{27}$$

$$\omega_b^2 + \omega_a^2 = 2\omega_0^2/k_1^2 \tag{28}$$

From these two equations we can find $k_1$ and $\omega_0$:

$$\frac{1}{\omega_0^2} = \frac{1}{2}\left(\frac{1}{\omega_a^2} + \frac{1}{\omega_b^2}\right) \tag{29}$$

$$\frac{1}{k_1} = \frac{1}{2}\left(\frac{\omega_b}{\omega_a} + \frac{\omega_a}{\omega_b}\right) \tag{30}$$

From equations (23) and (24)

$$Z_{i1}^2 = -(\omega^2 k_1^2/C_1^2)(\omega^4 k_1^2 - 2\omega^2\omega_0^2 + \omega_0^4)^{-1}$$

The form of $Z_{i1}$ in the pass band is shown in Fig. 24

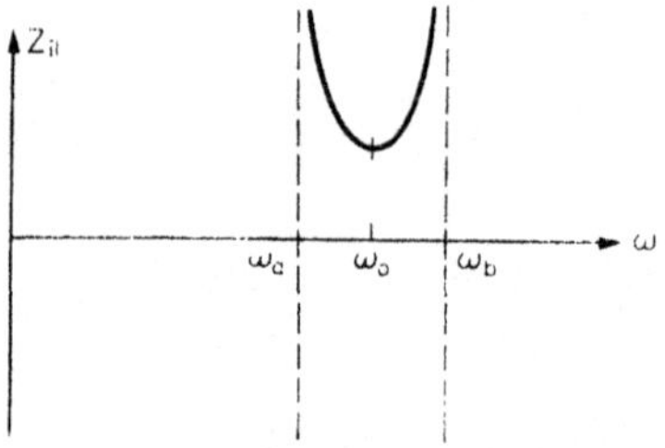


FIG. 24. *Image impedance in pass band (schematic)*

The minimum is somewhere near $\omega_0$, and at this frequency

$$Z_{i1} = R_1 = \frac{k_1}{k}\frac{1}{\omega_0 C_1} \tag{31}$$

Similarly at the same frequency

$$Z_{i2} = R_2 = \frac{k_1}{k}\frac{1}{\omega_0 C_2} \tag{32}$$

Equations (29)–(32) may be used for design purposes: choice of $\omega_a$ and $\omega_b$ gives $\omega_0$ and $k$. If, as is often the case, $C_1$ and $C_2$ are given $R_1$ and $R_2$ may then be found. If they are equated to source and load impedances, a single peak curve will be obtained. By increasing the terminating values slightly a double peak will be obtained.

Although this method does not give a prescribed response curve, it has the great advantage of simply giving component values which are not far wrong, and is as easy to apply for wide bands as for narrow bands. Small adjustments are always necessary in practical circuits, for example because $C_1$ and $C_2$ are not exactly known, and these will usually secure the desired response.

## DISSIPATION IN FILTERS

The effects of lossy inductors and capacitors cannot be simply allowed for in the methods of filter design outlined in this chapter. There is no fundamental difficulty in calculating the propagation constant with resistive as well as reactive elements, but the sharply defined mathematical difference between pass and stop bands immediately disappears.

An approach is given in Chapter 8 which applies when losses are incorporated uniformly into all reactive elements. The change in properties can then be simply estimated.

### *Examples on Chapter 7*

1. Perform an $m_T$-derivation on the band-pass section of Fig. 2 (assuming $Z_1 Z_2 = R_0^2$). Show that the resulting shunt arm may

be replaced by one consisting of two series $L$–$C$ circuits in parallel.

2. Show that the component values in the sections below may be chosen so that the T-image impedance is the same as that of Fig. 2.

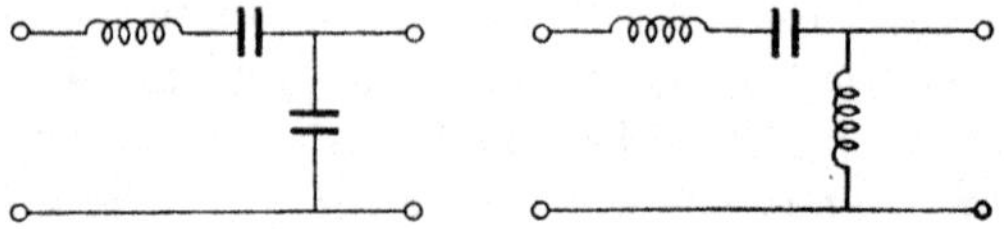

3. Show that the component values in the sections below may be chosen so that the Π-image impedance is the same as that of Fig. 2:

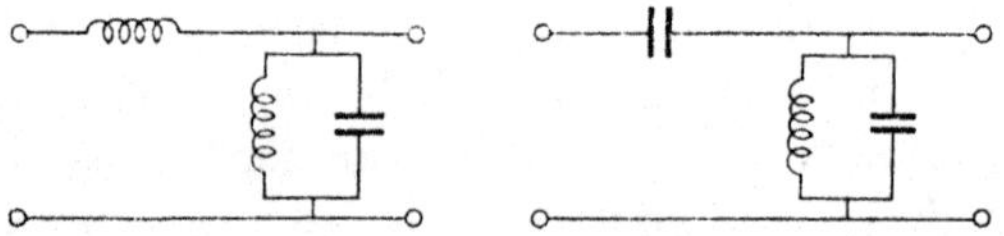

4. Investigate the filter properties of the section shown below. Show that for a single pass band the relation $(1-k^2)L_2 C_2 = L_1 C_1$ must be satisfied. When this is so, obtain expressions for the circuit elements in terms of the cut-off frequencies $\omega_1$, $\omega_2$ and the image impedances $R_1$, $R_2$ at the frequency $\sqrt{(\omega_1 \omega_2)}$.

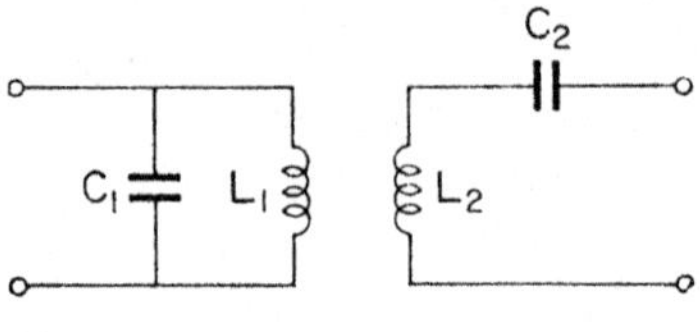


$$M = k\sqrt{L_1 L_2}$$

*Answer:*

$$k = (\omega_2-\omega_1)/(\omega_2^2+\omega_1^2-\omega_1\omega_2)^{1/2}$$
$$L_1 = (\omega_2-\omega_1)R_1/\omega_1\omega_2$$
$$C_1 = 1/R_1(\omega_2-\omega_1)$$
$$L_2 = R_2/(\omega_2-\omega_1)(1-k^2)$$
$$C_2 = (\omega_2-\omega_1)/R_2\omega_1\omega_2$$

5. Show that the admittance of a piezo-electric crystal, as specified by the equivalent circuit shown in Fig. 13 with $R = 0$, may be expressed in the form $j\omega C_0(\omega_2^2 - \omega^2)/(\omega_1^2 - \omega^2)$ with $\omega_2 > \omega_1$.

Four paired crystals in a lattice section as shown in Fig. 20 have parameters $C_{01}$, $\omega_1$, $\omega_2$ and $C_{02}$, $\omega_3$, $\omega_4$ respectively. Show that for a single pass band $\omega_2 = \omega_3$, that the image impedance at the frequency $\sqrt{(\omega_1 \omega_4)}$ is $1/[\omega_4 \sqrt{(C_{01} C_{02})}]$, and that for infinite attenuation to occur at frequencies $\omega_\infty$, $\omega_1 \omega_4/\omega_\infty$, $\omega_2^2 = \omega_1 \omega_4$ and

$$\frac{C_{02}}{C_{01}} = \frac{(\omega_2^2 - \omega_\infty^2)^2}{(\omega_4^2 - \omega_\infty^2)(\omega_1^2 - \omega_\infty^2)}$$

# 8 NETWORK FUNCTIONS

THE purpose of this chapter is to give a brief introduction to the theory of network functions, which forms the basis of network synthesis. It concerns the algebraic form of possible network functions regarded as functions of frequency, or more conveniently of the variable $p = j\omega$.

## THE DETERMINANT $\Delta$

We may consider firstly the determinant of the nodal admittance matrix.

$$\Delta = \begin{vmatrix} P_{11} & P_{12} & . & . & . & P_{1,n-1} \\ P_{21} & P_{22} & . & . & . & P_{2,n-1} \\ . & . & . & . & . & . \\ P_{n-1,1} & P_{n-1,2} & . & . & . & P_{n-1,n-1} \end{vmatrix} \tag{1}$$

In this determinant the elements $P_{ij}$ are all of the form

$$P_{ij} = pC_{ij} + 1/(pL_{ij}) + 1/R_{ij} \tag{2}$$

in which the coefficients of $p$ are real numbers. Bearing in mind the definition of the determinant,

$$\Delta = \sum(\pm) P_{1\alpha} P_{2\beta} \ldots P_{n-1,\nu}$$

given in Chapter 2, when all coefficients in equation (2) are non-zero we may write $\Delta$ in the form

$$\Delta = \frac{1}{p^{n-1}} \times \text{polynomial of degree } 2(n-1) \text{ in } p$$

$$= \frac{1}{p^{n-1}}(\alpha_{2n-2}\, p^{2n-2} + \alpha_{2n-3}\, p^{2n-3} + \ldots + \alpha_1\, p + \alpha_0) \qquad (3)$$

Although in particular cases some of the coefficients $\alpha_i$ may vanish, $\Delta$ can always be expressed in the form of equation (3), as a polynomial in $p$ with real coefficients divided by some power of $p$.

The other determinants which are associated with the solutions previously given are cofactors of $\Delta$, i.e. are formed from $\Delta$ by deletion of one row and one column. It therefore follows that they have the same algebraic form. For example,

$$\Delta_{11} = \begin{vmatrix} P_{22} & P_{23} & \cdot & \cdot & \cdot & P_{2,n-1} \\ P_{32} & P_{33} & \cdot & \cdot & \cdot & P_{3,n-1} \\ \cdot & \cdot & \cdot & \cdot & \cdot & \cdot \\ P_{n-1,1} & P_{n-1,2} & \cdot & \cdot & \cdot & P_{n-1,n-1} \end{vmatrix}$$

$$= \frac{1}{p^{n-2}}(\beta_{2n-4}\, p^{2n-4} + \beta_{2n-5}\, p^{2n-5} + \ldots + \beta_1\, p + \beta_0) \qquad (4)$$

Again, in particular cases some of the coefficients may vanish. It must be noted that the coefficients $\beta_i$ are formed from the same variables as are included in $\alpha_i$, and the vanishing of some of the $\alpha_i$ may correspond to the vanishing of certain $\beta_i$. Hence the actual degree of the polynomials of equations (3) and (4) will always be related. It follows from the reality of the coefficients of $p$ in equations (3) and (4) that the equations

$$\Delta = 0, \qquad \Delta_{11} = 0$$

will have for roots either real values of $p$ or complex values occurring in conjugate complex pairs.

Substitution of the expression for $\Delta$ and $\Delta_{11}$ in equation (3) of Chapter 3 gives the general form of an impedance function

$$Z = \frac{\Delta_{11}}{\Delta} = p \frac{\sum_{i=1}^{2n-4} \beta_i p^i}{\sum_{i=1}^{2n-2} \alpha_i p^i} \tag{5}$$

Similar expressions apply to the elements of the impedance matrix of a 2-port network defined by equations (6) and (7) of Chapter 3.

The investigation by algebraic methods of the restrictions placed on such coefficients as $\alpha_i$, $\beta_i$ by realisability, passivity, etc. of a network, the determination of necessary and sufficient conditions that a function shall for example be a physically realisable impedance function and similar problems are proper to the study of network synthesis, which is outside the scope of this book. However certain properties for restricted classes of networks may be obtained without use of these algebraic methods by appealing to transient properties of networks. These will now be considered.

## THE FORM OF TRANSIENT SOLUTION OF A NETWORK

It has been pointed out earlier (Chapter 2) that the general network equations are differential equations. These arise from the branch equations of the type

$$\left.\begin{aligned} u &= L\,\mathrm{d}i/\mathrm{d}t \\ i &= C\,\mathrm{d}u/\mathrm{d}t \end{aligned}\right\} \tag{6}$$

By reference to Chapter 2 it will be obvious that the occurrence of $j\omega$, or $p$, in an expression is mathematically the result of having a differentiation to perform, leading to impedance relations from equation (6)

$$\left.\begin{aligned} U &= j\omega LI \\ I &= j\omega CU \end{aligned}\right\} \tag{7}$$

We may therefore reconstruct the differential equations from the steady-state equations by replacing $p$ by the operation $\mathrm{d}/\mathrm{d}t$. The nodal equations may be written

$$p\sum_{s=1}^{n-1} P_{rs}V_s = pI_r, \qquad r = 1, 2, \ldots, n-1$$

If we replace $p$ by $\mathrm{d}/\mathrm{d}t$ we have in place of $p\,P_{rs}V_s$

$$\frac{\mathrm{d}^2}{\mathrm{d}t^2}(C_{rs}v_s)+\frac{\mathrm{d}}{\mathrm{d}t}\left(\frac{v_s}{R_s}\right)+\frac{v_s}{L_{rs}} \tag{8}$$

Hence the set of differential equations takes the form

$$\sum_{s=1}^{n-1}\left(C_{rs}\frac{\mathrm{d}^2v_s}{\mathrm{d}t^2}+\frac{1}{R_{rs}}\frac{\mathrm{d}v_s}{\mathrm{d}t}+\frac{v_s}{L_{rs}}\right)=\frac{\mathrm{d}i_r}{\mathrm{d}t}, \quad r = 1, 2, \ldots, n-1 \tag{9}$$

In general, we introduce unnecessary complications by multiplying each equation by $p$, since a particular equation may already be in differential form because of the absence of the $1/L_{rs}$ terms. This point is of considerable importance for some aspects of the theory, but does not cause confusion in the present context.

The complete solution of the set of equations (9) is made up of particular integrals and a complementary function involving a certain number of arbitrary constants. This set of equations is of the type "linear with constant coefficients" and may be tackled by elementary methods.

## THE COMPLEMENTARY FUNCTIONS

The complementary functions satisfy equations (9) with zero on the right-hand side. Electrically they correspond to the behaviour of the network after all driving currents are removed.

The equations to be solved may be written

$$\sum_{s=1}^{n-1}\left(C_{rs}D^2+\frac{1}{R_{rs}}D+\frac{1}{L_{rs}}\right)v_s = 0 \tag{10}$$

where $D \equiv \mathrm{d}/\mathrm{d}t$. By the standard algebraic procedure for solving linear simultaneous algebraic equations, we may eliminate all variables save one. This process will yield the same equation in each case, which may be written

$$\Delta'(D)v_s = 0 \qquad s = 1, 2, \ldots, n-1 \tag{11}$$

where $\Delta'(D)$ is a polynomial in $D$ given by the determinant

$$\Delta'(D) = \left| C_{rs} D^2 + \frac{1}{R_{rs}} D + \frac{1}{L_{rs}} \right| \tag{12}$$

Equation (11) is also a linear differential equation with constant coefficients. The standard form of solution is

$$v_s = \lambda_s e^{\mu t} \tag{13}$$

where $\mu$ must satisfy

$$\Delta'(\mu) = 0 \tag{14}$$

It is easily seen that this equation is also

$$\mu^{n-1} \Delta(\mu) = 0$$

where $\Delta(\mu)$ is the determinant of equation (1) with $p$ replaced by $\mu$. Hence the values of $\mu$ in equation (13) satisfy

$$\Delta(\mu) = 0 \tag{15}$$

We thus draw the conclusion that, provided no repeated factors exist,

$$v_s = \sum_{r=1}^{k} \lambda_{sr} \exp(\mu_r t) \tag{16}$$

The actual number of constants will depend on the degree of $\Delta(\mu)$. The constants for different values of $s$ will also be interrelated.

With the same nomenclature equation (15) gives

$$\Delta(p) = \frac{K}{p^{n-1}} (p-\mu_1)(p-\mu_2)\ldots(p-\mu_k) \tag{17}$$

If repeated factors exist, a typical term in (11) is of the type

$$(\Delta - \mu_r)^q v_s = 0$$

This has a solution

$$v_s = (a_{s1} + a_{s2}\, t + \ldots + a_{s,q-1}\, t^{q-1}) \exp(\mu_r\, t) \tag{18}$$

Equation (17) will have a term $(p - \mu_r)^q$.

## APPLICATION TO STABLE NETWORKS

A stable linear network is defined as one in which no voltage or current increases indefinitely with time in response to an initial disturbance lasting for a finite time only. A passive network is obviously a particular case of this category. The necessary and sufficient condition for this to be true is that no term in the complementary function shall increase indefinitely with time, and hence that no possible value of $\mu$ satisfying equation (15) shall have a positive real part. Hence

$$\mathrm{Re}(\mu_r) \leqq 0, \qquad r = 1, 2, \ldots, k \tag{19}$$

The case of a repeated root of equation (15) which is purely imaginary requires special consideration. At first sight equation (18) implies that such a root would give increasing voltages and is not therefore possible. This argument is however unsound. Since we are concerned with a set of linear equations and equation (18) applies for all $s$, the constants are interrelated. In the general case for a $q$-fold repeated root there must be a total $q$ arbitrary constants associated with that root in the expressions for $v_s$, and one possible case yielding non-increasing solutions is $a_{s2} = a_{s3} = \ldots = a_{sq-1} = 0$. The $q$ arbitrary constants can be taken up in the different constants $a_{s1}$. It may be shown that this is a possible case, and that a multiple root of order $q$ of $\Delta(\mu) = 0$ can happen. All the cofactors of the determinant $\Delta(\mu)$ will then have zeros at this value of $\mu$ of at least $(q-1)$-fold multiplicity.

## ZEROS OF THE COFACTORS OF $\Delta$

We may make similar conclusions about the symmetrical cofactor $\Delta_{ii}$. If we form from our network a new network by adding a conductance between node $i$ and the reference node $n$, the nodal admittance matrix of the new network differs from the old only by addition to the term $P_{ii}$. If the added conductance is increased indefinitely, i.e. becomes a short circuit, the term $P_{ii}$ also increases indefinitely, and the determinant approaches a value

$$P_{ii}\Delta_{ii} \tag{20}$$

Similarly the cofactor $\Delta_{rr}$ of the modified network approaches the value

$$P_{ii}\Delta_{rrii}$$

where $\Delta_{rrii}$ is the cofactor of $\Delta$ obtained by deleting rows $i$ and $r$ and columns $i$ and $r$. Hence $\Delta_{ii}$ is the determinant $\Delta$ for the network formed from the original network by shorting node $i$ to the reference node. If the resulting network is stable, the conditions deduced above for the zeros of $\Delta$ apply also to the zeros of $\Delta_{ii}$ and $\Delta_{rrii}$. These arguments cannot be applied to non-symmetrical cofactor $\Delta_{ij}$, $i \neq j$, which can in fact have zeros with real positive parts.

## SUMMARY OF PROPERTIES OF NETWORK DETERMINANTS

(i) Each determinant may be expressed as a polynomial in $p$ with real coefficients divided by some power of $p$. Hence

(ii) the zeros are real or occur in conjugate complex pairs and

(iii) the determinant $\Delta(j\omega)$ has a real part which is an even function of $\omega$ and an imaginary part which is an odd function.

(iv) for a stable network no zeros of $\Delta(p)$ have positive real parts.

(v) if a multiple zero of $\Delta$ occurs for a purely imaginary value of $p$, all cofactors of $\Delta$ have the same zero to the appropriate multiplicity.

All these conditions apply to cofactors of the type $\Delta_{ii}$ but only (i), (ii) and (iii) apply to those of type $\Delta_{ij}$, $i \neq j$.

## IMPEDANCE FUNCTIONS

The general expression for an impedance function was given in equation (5). We may now add to this the condition that the zeros of numerator and denominator polynomials obey the conditions given above as regards position and multiplicity. In this equation, which contains all the terms only if all components are present, $Z \to 1/p$ as $p \to \infty$. This is to be expected since a shunt capacity will appear at the input. In general it may be shown that $Z$ tends either to $p^{\pm 1}$ or to a constant as $p \to \infty$. This only states that for large $p$ we have at most an effectively inductive, capacitative or resistive network depending on configuration. For large frequencies the impedance can never increase faster than an inductance or decrease more rapidly than a capacity.

A particular case arises when the impedance is that of a network composed of reactances only. In this case equation (2) becomes

$$P_{ij} = pC_{ij} + 1/(pL_{ij})$$

and hence equation (3) becomes

$$\frac{1}{p^{n-1}} \times \text{polynomial of degree } n-1 \text{ in } p^2$$

There is no mechanism for energy dissipation in a reactive network so that the transient solution will contain only exponential terms of the type $\exp(j\omega_r t)$. The zeros of $\Delta(p)$ and $\Delta_{11}(p)$ will be imaginary. The expression for $Z$ given by equation (5) will therefore become

$$Z = p\frac{\prod_{r=1}^{n-2}(p^2+\omega_r^2)}{\prod_{r=1}^{n-1}(p^2+\Omega_r^2)}$$

both $\omega_r$ and $\Omega_r$ being real. No repeated factors are possible because of the relations between the multiplicity of the roots of $\Delta$ and $\Delta_{11}$.

It will be shown in Chapter 9 by consideration of energy that the rate of change of reactance with frequency is always positive. Hence the expression for reactance becomes

$$X(\omega) = \omega\frac{(\omega_1^2-\omega^2)(\omega_2^2-\omega^2)\ldots(\omega_{n-2}^2-\omega^2)}{(\Omega_1^2-\omega^2)(\Omega_2^2-\omega^2)\ldots(\Omega_{n-1}^2-\omega^2)}$$

with $\qquad \Omega_1 < \omega_1 < \Omega_2 < \omega_2 < \ldots < \omega_{n-2} < \Omega_{n-1}$

This latter condition is necessary to ensure $\mathrm{d}X/\mathrm{d}\omega > 0$.

## TRANSFER FUNCTIONS

A transfer function is one which relates conditions at one port of a 2-port to those at the other. If, for example, a 2-port is fed from a constant current generator at end 1, the open-circuit voltage at end 2 is

$$V_2 = z_{21} I_1 \tag{21}$$

We have seen that $z_{21}$ involves expressions such as

$$\Delta_{21}/\Delta$$

We can say, as previously, that this is a ratio of polynomials, but only for the denominator can we restrict the position of those zeros. Neither can we relate the degree of the numerator to that of the denominator.

The possibility that the zeros of $z_{21}$ can have positive real parts enables a distinction to be made between networks for which the zeros are restricted to have negative real parts and those which are not. The distinction is an important one because it shows that it is possible to modify the phase response of a network without modifying its gain response.

If we consider a 2-terminal plus reference network equation (21) becomes

$$V_2 = \frac{\Delta_{21}}{\Delta} I_1 = F(p) I_1 \tag{22}$$

Consider the transfer function defined by

$$F'(p) = \frac{\Delta_{21}}{\Delta} \frac{(p-p_1)(p-p_1^*)}{(p+p_1)(p+p_1^*)} \tag{23}$$

in which $$\text{Re}(p_1) > 0$$

We may interpret this as $\Delta'_{21}/\Delta'$, $\Delta'$ being a network determinant which has zeros at $-p_1$, $-p_1{}^*$ in addition to the zeros of $\Delta$, and $\Delta'_{21}$ a cofactor of $\Delta'$ which has zeros at $p_1$, $p_1{}^*$ in addition to those of $\Delta_{21}$. The additional zeros do not violate the conditions necessary to stability.

Placing $p = j\omega$

$$|F'(j\omega)| = |F(j\omega)| \tag{24}$$

since $$\left| \frac{(j\omega - p_1)(j\omega - p_1^*)}{(j\omega + p_1)(j\omega + p_1^*)} \right| = 1$$

Hence the networks whose transfer functions are $F$ and $F'$ have both the same magnitude of response at any frequency, but have different phases.

A network for which the transfer function has no zeros with positive real parts is called a minimum-phase network. If the transfer function has such zeros it is called a non-minimum-phase network. The name is a little unfortunate since it gives the idea that the phase of $F'(j\omega)$ is greater than that of $F(j\omega)$, which is not necessarily true.

## THE PHASE SHIFT OF AN ALL-PASS NETWORK

A network for which the transfer function has unity modulus is called an all-pass section. We have already had an example in Chapter 7, the simple terminated lattice with inductance and capacity arms. The additional phase shift of $F'(j\omega)$ is the phase of the function

$$\frac{(j\omega-\alpha-j\beta)(j\omega-\alpha+j\beta)}{(j\omega+\alpha+j\beta)(j\omega+\alpha-j\beta)} \tag{25}$$

in which $\alpha+j\beta=p_1$, and $\alpha,\beta>0$. On rewriting, expression (25) becomes

$$\frac{\alpha-j(\beta+\omega)}{\alpha+j(\beta+\omega)}\cdot\frac{\alpha-j(\omega-\beta)}{\alpha+j(\omega-\beta)}$$

Hence the phase angle $\phi$ is given by

$$\phi = -2\left[\tan^{-1}\left(\frac{\omega+\beta}{\alpha}\right)+\tan^{-1}\left(\frac{\omega-\beta}{\alpha}\right)\right] \tag{26}$$

the magnitude of both inverse tangents being regarded as less than $\pi/2$. This makes $\phi(-\omega)=\phi(\omega)$. This function is shown graphically in Fig. 1.

The effect of adding an all-pass network is therefore to decrease the phase algebraically, although in magnitude the effect may be to give a greater or lesser total phase at any one frequency.

## NETWORK CONFIGURATIONS FOR NON-MINIMUM-PHASE PROPERTIES

It is not, in general, possible to say from the configuration whether a network will have non-minimum-phase properties or not. It can however be said that a ladder network is necessarily minimum-phase. This is because any zero of a transfer function in such a network can only occur through either a series impedance becoming infinite or a shunt impedance becoming zero. From the properties of impedance functions we know that the zeros and infinities of such functions must necessarily have real parts negative, and hence no zero of the transfer function can have a positive real part.

In general non-minimum-phase properties can be expected from bridge-type networks.

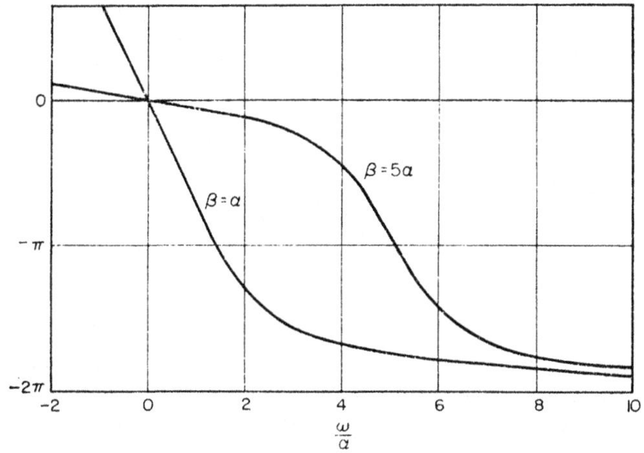


FIG. 1. *The function* $-2\left[\tan^{-1}\frac{\omega-\beta}{\alpha}+\tan^{-1}\frac{\omega+\beta}{\alpha}\right]$

## UNIFORM DISSIPATION

The whole question of the effect of using real, and hence lossy, elements in filter circuits designed for reactive elements has been left unanswered. A partial answer can be given with the aid of the theory of the present chapter.

It has been pointed out that a network function is an algebraic function of the variable $p$ with real coefficients. If we replace $p$ by $p+\varepsilon$ in the algebraic expression for a transfer function, this is equivalent to replacing the expressions $Lp$, $Cp$ for the impedance and admittance of each inductance and capacity by $L(p+\varepsilon)$, $C(p+\varepsilon)$ respectively. This mathematical replacement is equivalent to the circuit replacement shown in Fig. 2.

Suppose that for a particular network one of the network functions (e.g. impedance, or transfer function) is $H(p)$. Then the function

$$H_1(p) = H(p+\varepsilon) \tag{27}$$

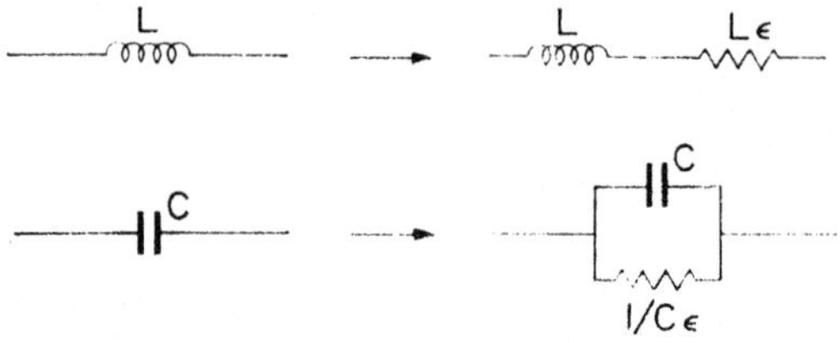


FIG. 2. *Uniform dissipation*

is the same network function for a network obtained from the original network by adding in series with each inductance $L$ a resistance of $\varepsilon L$ and in parallel with each capacity $C$ a resistance $1/(\varepsilon C)$. This result holds whatever the original network. If the original one is reactive, the addition of resistance in this way is said to give uniform dissipation.

Equation (27) enables $H_1(p)$ to be calculated when the form of $H(p)$ is known. If $\varepsilon$ is small, then above a certain minimum frequency $\varepsilon \ll |p|$. When this condition is satisfied we may expand $H(p+\varepsilon)$ as a Taylor series:

$$H(p+\varepsilon) = H(p) + \frac{\varepsilon}{1!}\frac{\mathrm{d}H}{\mathrm{d}p} + \frac{\varepsilon^2}{2!}\frac{\mathrm{d}^2H}{\mathrm{d}p^2} + \dots \tag{28}$$

Hence for small $\varepsilon$ and for values of $p$ where the series (28) is rapidly convergent, we have

$$H_1(p) = H(p+\varepsilon) \simeq H(p) + \varepsilon\frac{\mathrm{d}H}{\mathrm{d}p}$$

Because of the nature of functions of a complex variable we have

$$H_1(j\omega) \simeq H(j\omega)+\varepsilon\frac{\mathrm{d}}{\mathrm{d}(j\omega)}H(j\omega) \tag{29}$$

We may take for $H(p)$ any network function. Take then the logarithm of the transfer function. Then

$$H(j\omega) = \log_e F(j\omega) = \alpha(\omega)+j\beta(\omega) \tag{30}$$

$\alpha$ is the logarithmic gain and $\beta$ is the phase. Substituting this expression in equation (29) we find

$$\left.\begin{aligned}\alpha_1(\omega)-\alpha(\omega) &\simeq \varepsilon\,\mathrm{d}\beta/\mathrm{d}\omega\\ \beta_1(\omega)-\beta(\omega) &\simeq -\varepsilon\,\mathrm{d}\alpha/\mathrm{d}\omega\end{aligned}\right\} \tag{31}$$

These equations will not apply at frequencies when $\alpha$ and $\beta$ are changing rapidly. They may however be used to give a reasonable estimate over much of the range of frequencies.

## *Examples on Chapter 8*

1. Obtain expressions for the determinants of the nodal admittance matrices of the following networks as rational functions of $p$. Confirm that the differential equation for the current in each has the appropriate complementary function.

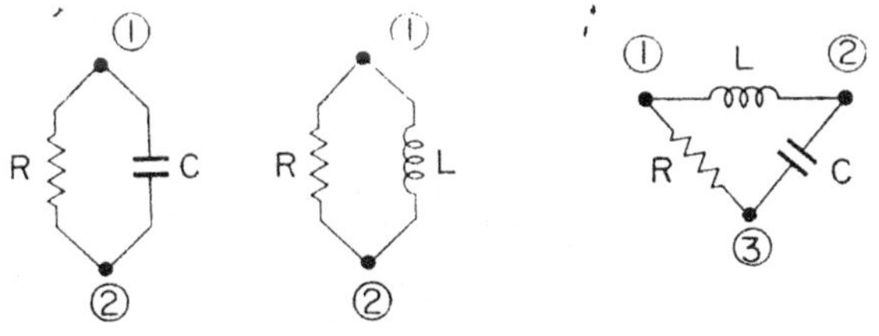


*Answer:* $G+pC$, $(R+pL)^{-1}$ $G(p^2LC+pC/G+1)/pL$

2. Obtain expressions in the form of rational algebraic functions for the impedances of the following circuits, and hence show that members of each pair can have identical impedances.

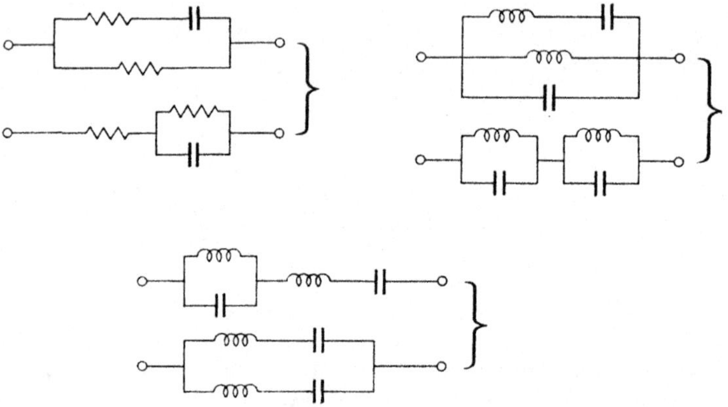

3. Obtain expressions for the determinant, $\Delta$, of the nodal admittance matrix and the cofactors $\Delta_{11}$ and $\Delta_{13}$ for the network below. Investigate the zeros of $\Delta$, $\Delta_{11}$, and $\Delta_{13}$. Show that the complex zeros of $\Delta_{13}$ have positive real parts. For the case $\alpha = 12$, solve $\Delta_{13} = 0$ and show that the roots of $\Delta_{11} = 0$ lie near the values $-12$, $\pm j$.

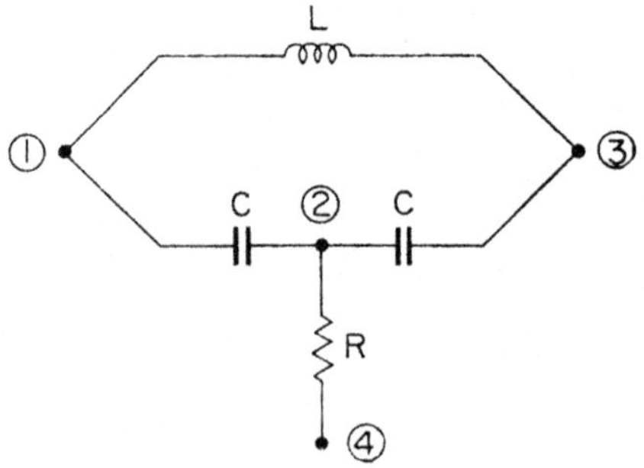

*Answer:* Writing $x = p\sqrt{(LC)}$

$$\Delta = (C/RL)(2+x^2); \quad \Delta_{11} = (C/Lx)(x^3+\alpha x^2+2x+\alpha)$$

$$\Delta_{13} = (C/Lx)(x^3+2x+\alpha)$$

$$x = -2, \quad 1 \pm j\sqrt{5}$$

4. Write down the nodal admittance matrix for the network below. Evaluate $\Delta$ and $\Delta_{11}$ and hence obtain an expression for the input impedance. Show that the zeros of the numerator and denominator expressions interlace.

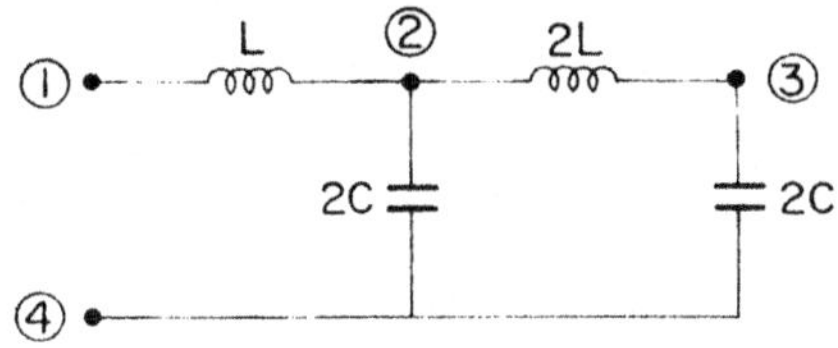


*Answer:* Writing $x = p\sqrt{(LC)}$

$$\Delta = 2\left(\frac{C}{L}\right)^{\frac{1}{2}} \frac{2x^2+1}{x}, \qquad \Delta_{11} = \frac{C}{L}\,\frac{8x^4+8x^2+1}{2x^2}$$

5. A terminated balanced lattice section is shown below, the impedances being such that $\zeta_1\zeta_2 = R_0^2$. Find the ratio $V_2/V_1$ and show that the network is not of the minimum-phase type if $R > R_0$.

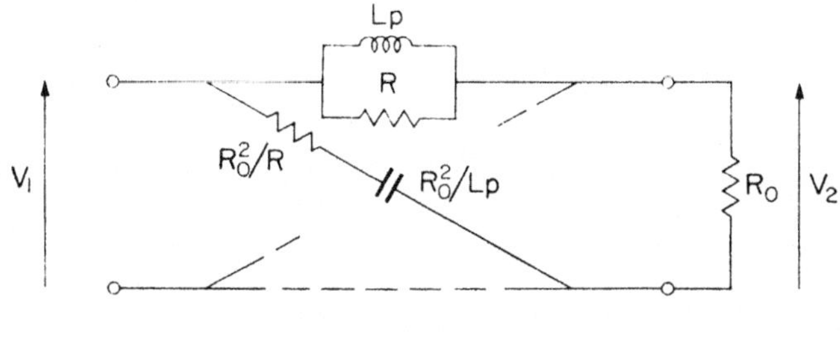

6. Show that if in the previous question $R > R_0^2$, the lattice section is equivalent to the cascaded sections given in the figure below, one of which is minimum-phase and the other all-pass.

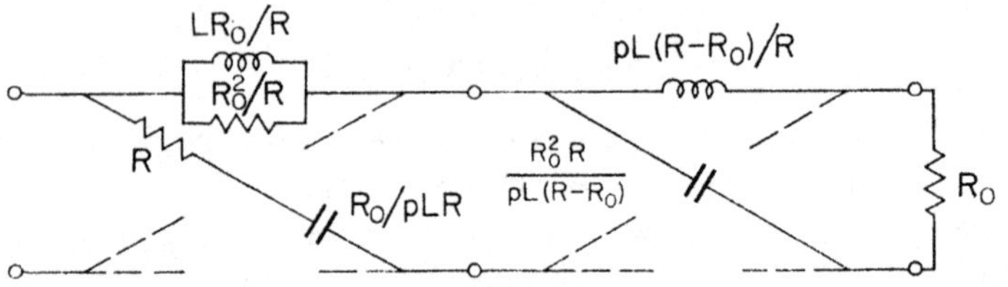


7. Show that for a purely reactive impedance the effect of introducing a small uniform dissipation $\varepsilon$ is to add a resistance $\varepsilon\,\mathrm{d}X/\mathrm{d}\omega$ in series, or a conductance $\varepsilon\,\mathrm{d}B/\mathrm{d}\omega$ in parallel, $X$ and $B$ being respectively reactance and susceptance. Apply these formulae to the case of a series $L$–$C$ circuit, and compare the results with those obtained from exact analysis.

8. The image propagation constant for a low-pass filter of unit cut-off frequency is given by the expression

$$\exp 2T = \frac{\sqrt{(1+p^2)}+p}{\sqrt{(1+p^2)}-p}$$

Show that $\mathrm{d}T/\mathrm{d}p = (1+p^2)^{-\frac{1}{2}}$ and hence derive an approximate expression for the attenuation in the pass band on adding uniform dissipation $\varepsilon$. Check the actual behaviour for $p$ small and for $p$ near unity, assuming $\varepsilon \ll 1$.

*Answer:* $\alpha = \varepsilon(1-\omega^2)^{-\frac{1}{2}}$, $\omega$ not near unity. At $\omega = 1$, $\alpha \simeq \sqrt{\varepsilon}$.

# 9 ENERGY RELATIONS IN PASSIVE RECIPROCAL NETWORKS

IT is the purpose of this chapter to derive certain relations between the impedance description of a network and the energy in the network.

## INSTANTANEOUS RELATIONS

It is convenient for present purposes to deal with a series branch of a general network as shown in the figure.

The differential equation relating the variables is

$$e_r + u_r = L_r \frac{\mathrm{d}i_r}{\mathrm{d}t} + R_r i_r + \frac{q_r}{C_r}, \quad r = 1, 2, \ldots, b \tag{1}$$

where $$q_r = \int_{-\infty}^{t} i_r \, \mathrm{d}t$$

Multiplying equation (1) by $i_r$ we find

$$e_r i_r = -u_r i_r + \frac{\mathrm{d}}{\mathrm{d}t}\left(\frac{1}{2} L_r i_r^2\right) + i_r^2 R_r + \frac{\mathrm{d}}{\mathrm{d}t}\left(\frac{1}{2}\frac{q_r^2}{C_r}\right) \tag{2}$$

The term on the left-hand side is the instantaneous power produced by the source. The second and fourth terms on the right-hand side are the rates of increase of stored energy, i.e. power absorption, by the inductance and the capacity respectively. The third term is the power dissipated in the resistance. Summing equations (2) over all the branches we have

$$\sum_{r=1}^{b} e_r i_r = -\sum_{r=1}^{b} u_r i_r + \frac{d}{dt}(\mathscr{E}_m + \mathscr{E}_e) + P \tag{3}$$

in which $\mathscr{E}_m$, $\mathscr{E}_e$ represent the total energies stored inductively and capacitatively respectively, and $P$ the total rate of dissipation of energy. Since these terms include all possible places at which

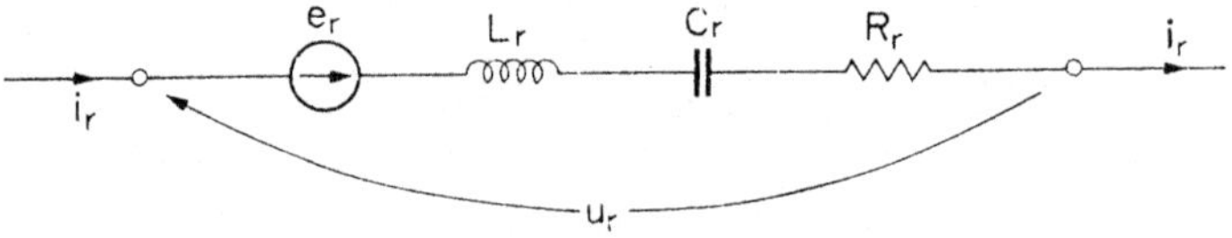


*The general branch*

energy can be extracted from the sources, and the left-hand side is the total energy supplied we must have

$$\sum_{r=1}^{b} u_r i_r = 0 \tag{4}$$

This is of course the mathematical statement that the constraints, the network connections, cannot dissipate power. This equation must be an identity, and depending not on the values of the circuit elements but on the interconnections. The equations restricting the variables $u_r$ and $i_r$ are equations (2) and (3) of Chapter 1:

$$\sum_{s=1}^{b} B_{rs} u_s = 0, \qquad r = 1, 2, \ldots, l \tag{5}$$

$$\sum_{s=1} A_{rs} i_s = 0, \qquad r = 1, 2, \ldots, n \tag{6}$$

The first relates to a set of independent loops, and the second to the nodes. Consider the nodal method of analysis: the definition of the coefficients $A_{rs}$ is $+1$ if the current $i_s$ flows away from node $r$, $-1$ if current $i_s$ flows into the node $r$, and zero if current $i_s$ is not connected to node $r$.

It follows immediately from this definition that the relation between branch voltages $u_r$ and nodal voltages $v_s$, referred to node $n$, is

$$u_r = \sum_{s=1}^{n-1} A_{sr} v_s \tag{7}$$

Hence the left-hand side of equation (4) is

$$\sum_{r=1}^{b} \sum_{s=1}^{n-1} A_{sr} v_s i_r = \sum_{s=1}^{n-1} v_s \sum_{r=1}^{b} A_{sr} i_r$$

This vanishes by equation (6), and hence we see that equation (4) is indeed satisfied identically. Similar results may be established by the loop method.

## STEADY-STATE EQUATIONS

Transforming the equations as in Chapter 2 we have for the sinusoidal steady-state

$$E_r = -U_r + (j\omega L_r + R_r + 1/j\omega C_r) I_r \tag{8}$$

$$U_r = \sum_{s=1}^{n-1} A_{sr} V_s \tag{9}$$

$$\sum_{s=1}^{b} A_{rs} I_s = 0 \tag{10}$$

Multiplying equation (8) by the complex conjugate of $I_r$ and summing

$$\sum_{r=1}^{b} E_r I_r^* = -\sum_{r=1}^{b} U_r I_r^* + \sum_{r=1}^{b} (j\omega L_r + R_r + 1/j\omega C_r) |I_r|^2 \tag{11}$$

The coefficients in equations (9) and (10) are real and hence $\sum_{r=1}^{b} U_r I^*$ vanishes identically. Further the remaining terms may be expressed in terms of energy.

The mean energy stored in $L_r$ is

$$\overline{\mathscr{E}_{mr}} = \tfrac{1}{2}L_r \overline{i_r^2} = \tfrac{1}{2}L_r |I_r|^2 \tag{12}$$

assuming that $I_r$ is calculated on an r.m.s. basis. For $C_r$ we have $i_r = dq_r/dt$ and hence $I_r = j\omega Q_r$. Therefore the mean energy stored in $C_r$ is

$$=\overline{\mathscr{E}_{er}} \frac{1}{2C_r}\overline{q_r^2} = \frac{1}{2C_r}|Q_r|^2 = \frac{1}{2C_r\omega^2}|I_r|^2 \tag{13}$$

Equation (11) then takes the form

$$\sum_{r=1}^{b} E_r I_r^* = \bar{P} + 2j\omega(\overline{\mathscr{E}_m} - \overline{\mathscr{E}_e}) \tag{14}$$

in which $\bar{P}$ = average rate of dissipation of energy

$\overline{\mathscr{E}_m}$ = mean energy stored inductively

$\overline{\mathscr{E}_e}$ = mean energy stored capacitatively.

## IMPEDANCE IN TERMS OF ENERGY

Consider the network with only one source, $E_1$. Equation (14) becomes

$$E_1 I_1^* = \bar{P} + 2j\omega(\overline{\mathscr{E}_m} - \overline{\mathscr{E}_e})$$

The impedance seen by the source is

$$Z = E_1/I_1$$

$$= \frac{1}{|I_1|^2}\{\bar{P} + 2j\omega(\overline{\mathscr{E}_m} - \overline{\mathscr{E}_e})\} \tag{15}$$

Alternatively, the admittance $Y$ is given by

$$Y = \frac{1}{|E_1|^2}\{\bar{P} - 2j\omega(\overline{\mathscr{E}_m} - \overline{\mathscr{E}_e})\} \tag{16}$$

These expressions show that the real part is related to the total loss in the network and the imaginary part to the difference between the energies stored in the two types of reactive components.

## MEAN TOTAL STORED ENERGY

The mean energy stored in branch $r$ is, according to equations (12) and (13)

$$\begin{aligned}\overline{\mathscr{E}_r} = \overline{\mathscr{E}_{mr}} + \overline{\mathscr{E}_{er}} &= \frac{1}{2}\left(L_r + \frac{1}{\omega^2 C_r}\right)|I_r|^2 \\ &= \frac{1}{2}\frac{\mathrm{d}X_r}{\mathrm{d}\omega}|I_r|^2\end{aligned} \tag{17}$$

Allowing the frequency to change whilst keeping $E_r$ fixed in magnitude and phase we have from equation (8)

$$0 = -\delta U_r + j\,\delta X_r I_r + (R_r + jX_r)\,\delta I_r$$

Hence, making use of equations (9) and (10) we find

$$0 = j\sum_{r=1}^{b} \delta X_r I_r I_r^* + \sum_{r=1}^{b} (R_r + jX_r) I_r^*\,\delta I_r \tag{18}$$

Taking the complex conjugate of equation (8), multiplying by $\delta I_r$ and summing we have

$$\sum_{r=1}^{b} E_r^*\,\delta I_r = \sum_{r=1}^{b} R_r I_r^*\,\delta I_r - j\sum_{r=1}^{b} X_r I_r^*\,\delta I_r \tag{19}$$

Adding equations (18) and (19) we find

$$\sum_{r=1}^{b} E_r^* \, \delta I_r = j \sum_{r=1}^{b} \delta X_r I_r I_r^* + 2 \sum_{r=1}^{b} R_r I_r^* \, \delta I_r$$

We hence obtain an expression for the total stored energy as

$$\overline{\mathscr{E}} = \frac{1}{2j}\left[\sum_{r=1}^{b} E_r^* \frac{\mathrm{d}I_r}{\mathrm{d}\omega} - 2 \sum_{r=1}^{b} R_r I_r^* \frac{\mathrm{d}I_r}{\mathrm{d}\omega}\right] \tag{20}$$

Even when only one driving voltage is considered it is not possible to reduce this expression to one involving only input current and voltage.

One special case is that of a purely reactive network, where $R_r = 0$ for all $r$. Equation (20) then becomes

$$\overline{\mathscr{E}} = \frac{1}{2j} \sum_{r=1}^{b} E_r^* \frac{\mathrm{d}I_r}{\mathrm{d}\omega} \tag{21}$$

If we further take only $E_1$ non-zero we may write

$$I_1 = jBE_1$$

in which $B$ is the input susceptance seen by $E_1$. Equation (21) then provides the relation

$$\overline{\mathscr{E}} = \tfrac{1}{2} |E_1|^2 \frac{\mathrm{d}B}{\mathrm{d}\omega}$$

whence

$$\frac{\mathrm{d}B}{\mathrm{d}\omega} = \frac{2\overline{\mathscr{E}}}{|E_1|^2} \tag{22}$$

In terms of input reactance

$$\frac{\mathrm{d}X}{\mathrm{d}\omega} = \frac{2\overline{\mathscr{E}}}{|I_1|^2} \tag{23}$$

We may combine these equations with (15) and (16) to obtain

$$\frac{1}{B}\frac{\mathrm{d}B}{\mathrm{d}\omega} = -\frac{1}{X}\frac{\mathrm{d}X}{\mathrm{d}\omega} = \frac{1}{\omega}\,\frac{\overline{\mathscr{E}_e} + \overline{\mathscr{E}_m}}{\overline{\mathscr{E}_e} - \overline{\mathscr{E}_m}} \tag{24}$$

Equations (22) and (23) show that reactance and susceptance functions have always positive slope. Equation (24) shows further that this slope is always greater than unity.

## GENERAL NETWORK DRIVEN BY A SINGLE SOURCE

If we restrict ourselves to this case, equation (20) can be expressed in more meaningful terms. We have

$$\bar{\mathscr{E}} = \tfrac{1}{2}\operatorname{Im}\left\{E_1^* \frac{\mathrm{d}I_1}{\mathrm{d}\omega}\right\} - \sum_{r=1}^{b} R_r \operatorname{Im}\left\{I_r^* \frac{\mathrm{d}I_r}{\mathrm{d}\omega}\right\} \tag{25}$$

With respect to the source $E_1$ let $I_r$ have phase angle $\theta_r$, giving

$$I_r = |I_r|\,\mathrm{e}^{j\theta_r}$$

Differentiating we find

$$\frac{\mathrm{d}I_r}{\mathrm{d}\omega} = \mathrm{e}^{j\theta_r}\frac{\mathrm{d}}{\mathrm{d}\omega}|I_r| + j|I_r|\,\mathrm{e}^{j\theta_r}\frac{\mathrm{d}\theta_r}{\mathrm{d}\omega}$$

Hence

$$I_r^* \frac{\mathrm{d}I_r}{\mathrm{d}\omega} = |I_r|\frac{\mathrm{d}}{\mathrm{d}\omega}|I_r| + j|I_r|^2\frac{\mathrm{d}\theta_r}{\mathrm{d}\omega}$$

Equation (25) may therefore be written

$$\bar{\mathscr{E}} = \tfrac{1}{2}\operatorname{Im}\left\{(E_1^* - 2R_1 I_1^*)\frac{\mathrm{d}I_1}{\mathrm{d}\omega}\right\} - \sum_{r=2}^{b} R_r |I_r|^2 \frac{\mathrm{d}\theta_r}{\mathrm{d}\omega} \tag{26}$$

Equation (26) provides a means of determining stored energy in terms of the behaviour of the currents in the resistors and in the input circuit. All branches need not contain resistive elements and the second term is only summed over those which do contain such elements. In such a determination $R_1$ is obviously in a different category to the others, since it is in series with the source and merely modifies the current flowing into the network. We may separate the effect of this component by considering our network to be a multi-port reactive network terminated in resistors at its output ports and fed at one port from a source of resistive impedance $R_1$. If the input port has impedance $Z_i$, presented to $E_1$ in series with $R_1$, we introduce the reflection coefficient

$$k = \frac{Z_i - R_1}{Z_i + R_1}$$

Now in terms of $Z_i$

$$E_1 = (Z_i + R_1)I_1 = R_1 I_1 \frac{2}{1-k}$$

Hence
$$I_1 = \frac{E_1}{2R_1}(1-k)$$

The first term of equation (26) then becomes

$$\tfrac{1}{2}\,\mathrm{Im}\left\{E_1^* k^* \frac{E_1}{2R_1}\left(-\frac{dk}{d\omega}\right)\right\}$$

If we denote $k = Ke^{j\phi}$ we have finally

$$\bar{\mathscr{E}} = -\frac{|E_1|^2}{4R_1}K^2\frac{d\phi}{d\omega} - \sum_{r=2}^{b} R_r |I_r|^2 \frac{d\theta_r}{d\omega} \tag{27}$$

A special case of (27) occurs when we consider a 2-port network fed from $R_1$ and fed into $R_2$. If it is also matched at the input we have

$$\bar{\mathscr{E}} = -R_2 |I_2|^2 \frac{d\theta_2}{d\omega}$$

giving an expression for rate of change of phase in terms of stored energy.

## RATE OF CHANGE OF THE MODULUS OF AN IMPEDANCE

It is possible to place an upper limit on $\left|\frac{dZ}{d\omega}\right|$. From equation (8) we have as before

$$0 = -\delta U_r + j\,\delta X_r I_r + (R_r + jX_r)\,\delta I_r$$

Hence, using equations (9) and (10)

$$0 = j\sum_{r=1}^{b} \delta X_r I_r^2 + \sum_{r=1}^{b} (R_r + jX_r) I_r \delta I_r \tag{28}$$

Also multiplying equation (8) by $\delta I_r$ and summing

$$\sum_{r=1}^{b} E_r \delta I_r = \sum_{r=1}^{b} (R_r + jX_r) I_r \delta I_r \tag{29}$$

Eliminating the term in $(R_r + jX_r)$ from equations (28) and (29)

$$\sum_{r=1}^{b} E_r \delta I_r = -j \sum_{r=1}^{b} \delta X_r I_r^2$$

Considering only a single source we have

$$\begin{aligned} E_1 \delta I_1 &= -j \sum_{r=1}^{b} \delta X_r I_r^2 \\ &= -2j\,\delta\omega \sum_{r=1}^{b} (\overline{\mathscr{E}_{er}} + \overline{\mathscr{E}_{mr}})\, e^{2j\theta_r} \end{aligned}$$

in which $\theta_r$ is the phase angle of $I_r$, as in the previous section.

Taking moduli of both sides

$$\begin{aligned} |E_1 \delta I_1| &= 2\,\delta\omega \left| \sum_{r=1}^{b} (\overline{\mathscr{E}_{er}} + \overline{\mathscr{E}_{mr}})\, e^{2j\theta_r} \right| \\ &\leqq 2\,\delta\omega \sum_{r=1}^{b} (\overline{\mathscr{E}_{er}} + \overline{\mathscr{E}_{mr}}) \\ &= 2\overline{\mathscr{E}}\,\delta\omega \end{aligned}$$

Hence
$$\left|\frac{dY}{d\omega}\right| \leq \frac{2\bar{\mathscr{E}}}{|E_1|^2} \tag{30}$$

The equality sign holds when $Y$ is a pure susceptance.

Similarly
$$\left|\frac{dZ}{d\omega}\right| \leq \frac{2\bar{\mathscr{E}}}{|I_1|^2}$$

*Acknowledgement:* the section on mean total stored energy and the following section owe much to a recent article by G. Kishi and K. Nakazawa: *I.E.E.E. Transactions on Circuit Theory*, Vol. CT-10, 1963, pp. 67–71.

# 10 MATRIX METHODS

THE results of Chapter 1 may be much more concisely presented using matrix notation. It is the purpose of this chapter to present those results in this form.

Consider first equation (3) of Chapter 1, repeated below

$$\sum_{s=1}^{b} A_{rs} i_s = 0, \qquad r = 1, 2, \ldots, n \tag{1}$$

These equations state that the total current flowing away from each node is zero. The value of $A_{rs}$ ($\pm 1$, 0) depends on whether or not the branch $s$ is connected to node $r$. We have defined

$A_{rs} = +1$ if branch $s$ is incident with node $r$ and directed away from it

$A_{rs} = -1$ if branch $s$ is incident with node $r$ and directed towards it

$A_{rs} = 0$ if branch $s$ is not incident with node $r$.

The matrix $\mathbf{A}_a$ is defined as the $n \times b$ matrix

$$\begin{bmatrix} A_{11} & A_{12} & . & . & . & A_{1b} \\ A_{21} & A_{22} & . & . & . & A_{2b} \\ . & . & . & . & . & . \\ A_{n1} & A_{n2} & . & . & . & A_{nb} \end{bmatrix}$$

Each column has one $+1$, one $-1$ and all the remaining terms zero.

If a column vector is formed from the branch currents

$$\mathbf{i} = \begin{bmatrix} i_1 \\ i_2 \\ \vdots \\ i_b \end{bmatrix}$$

equation (1) may be written

$$\mathbf{A}_a \mathbf{i} = 0 \tag{1'}$$

There are only $n-1$ independent rows in $\mathbf{A}_a$, since conservation of current at $n-1$ nodes ensures conservation at the remaining one. It is therefore convenient to delete one row from $\mathbf{A}_a$, say row $n$, and form the $n-1 \times b$ matrix $\mathbf{A}$

$$\mathbf{A} = \begin{bmatrix} A_{11} & A_{12} & \cdot & \cdot & \cdot & A_{1b} \\ A_{21} & A_{22} & \cdot & \cdot & \cdot & A_{2b} \\ \cdot & \cdot & \cdot & \cdot & \cdot & \cdot \\ A_{n-1,1} & A_{n-1,2} & \cdot & \cdot & \cdot & A_{n-1,b} \end{bmatrix}$$

This is termed the reduced node-branch incidence matrix, and the equation

$$\mathbf{A}\mathbf{i} = 0 \tag{2}$$

contains the useful equations from (1) and (1′).

Equation (2) of Chapter 1

$$\sum_{s=1}^{b} B_{rs} u_s = 0, \qquad r = 1, 2, \ldots, l \tag{3}$$

expressing the fact that there must be zero change in potential round any closed loop may be expressed similarly. From the definition of $B_{rs}$

$B_{rs} = +1$ if branch $s$ is incident with loop $r$ and directed with it

$B_{rs} = -1$ if branch $s$ is incident with loop $r$ and directed against it

$B_{rs} = 0$ if branch $s$ is not contained in loop $r$.

A column vector of the branch voltages may be formed:

$$\mathbf{u} = \begin{bmatrix} u_1 \\ u_2 \\ \vdots \\ u_b \end{bmatrix}$$

Equation (3) becomes

$$\mathbf{Bu} = 0 \qquad (3')$$

It is important to realise that $\mathbf{B}$ results from a choice of independent loops. Many sets of such loops may exist and $\mathbf{B}$ is not unique in the way in which $\mathbf{A}_a$ is unique.

## NODAL VOLTAGES

The relation between nodal voltages and branch voltages can be simply expressed in terms of the reduced-node-branch incident matrix $\mathbf{A}$. By considering the definition of the components of $\mathbf{A}$, it is seen that the branch voltages are given in terms of the nodal voltages by the matrix equation

$$\begin{bmatrix} u_1 \\ u_2 \\ \cdot \\ u_b \end{bmatrix} = \begin{bmatrix} A_{11} & A_{21} & \cdot & \cdot & \cdot & A_{n-1,1} \\ A_{12} & A_{22} & \cdot & \cdot & \cdot & A_{n-1,2} \\ \cdot & \cdot & \cdot & \cdot & \cdot & \cdot \\ A_{1b} & A_{2b} & \cdot & \cdot & \cdot & A_{n-1,b} \end{bmatrix} \times \begin{bmatrix} v_1 \\ v_2 \\ \cdot \\ v_{n-1} \end{bmatrix}$$

In this the nodal voltages are referred to node $n$ as origin, corresponding to the row which was deleted from $\mathbf{A}_a$ to form $\mathbf{A}$. The $b \times n$ matrix is formed from $\mathbf{A}$ by interchanging rows and columns. This operation is known as transposing and the transposed matrix is denoted by $\mathbf{A}^T$. The equation may then be written

$$\mathbf{u} = \mathbf{A}^T\mathbf{v} \qquad (4)$$

In formal nodal analysis it is customary to use the branch

notation of Fig. 4, Chapter 1 employing current generators. The relationship between branch voltage and current is

$$i_r - j_r = G_r u_r, \qquad r = 1, 2, \ldots, b \tag{5}$$

If a $b \times b$ matrix $\mathbf{Y}$ is introduced, defined as

$$\mathbf{Y} = \begin{bmatrix} G_1 & 0 & . & . & . & 0 \\ 0 & G_2 & . & . & . & 0 \\ . & . & . & . & . & . \\ 0 & 0 & . & . & . & G_b \end{bmatrix}$$

in which all the elements are zero except those on the diagonal, equations (5) may be written

$$\mathbf{i} - \mathbf{j} = \mathbf{Y}\mathbf{u} \tag{6}$$

Here $\mathbf{j}$ is a column vector formed from the currents $j_r$. Substituting from equation (4)

$$\mathbf{i} - \mathbf{j} = \mathbf{Y}\mathbf{A}^T\mathbf{v}$$

Multiplying by $\mathbf{A}$ and using equation (2)

$$-\mathbf{A}\mathbf{j} = \mathbf{A}\mathbf{Y}\mathbf{A}^T\mathbf{v} \tag{7}$$

The column vector $-\mathbf{A}\mathbf{j}$, which has $n-1$ elements is seen to be the column vector of the total currents flowing into nodes $1, 2, \ldots, n-1$, when all the nodes are joined together. It may therefore be identified with the column vector made of the currents $j_1^{(n)}, j_2^{(n)}, \ldots, j_{n-1}^{(n)}$ occurring in equation (7) of Chapter 1.

Comparison of the same equation with equation (7) then gives for the nodal admittance matrix $\mathbf{P}$ the expression

$$\mathbf{P} = \mathbf{A}\mathbf{Y}\mathbf{A}^T \tag{8}$$

## LOOP ANALYSIS

An exactly parallel set of results may be given. If loop currents $i_1^{(l)}, i_2^{(l)}, \ldots, i_l^{(l)}$ are taken as unknowns the branch currents are given by the equations

$$\mathbf{i} = \mathbf{B}^T\mathbf{i}^{(l)} \tag{9}$$

$\mathbf{B}^T$ being the transpose of $\mathbf{B}$.

Taking the branch configuration of Fig. 3, Chapter 1, using voltage generators

$$u_r + e_r = R_r i_r$$

or

$$\mathbf{u}+\mathbf{e} = \mathbf{Z}\mathbf{i} \tag{10}$$

in which $\mathbf{Z}$ is the diagonal matrix

$$\mathbf{Z} = \begin{bmatrix} R_1 & 0 & . & . & . & 0 \\ 0 & R_2 & . & . & . & 0 \\ . & . & . & . & . & . \\ 0 & 0 & . & . & . & R_b \end{bmatrix}$$

Hence

$$\mathbf{u}+\mathbf{e} = \mathbf{Z}\mathbf{B}^T\mathbf{i}^{(l)}$$

and

$$\mathbf{B}\mathbf{e} = \mathbf{B}\mathbf{Z}\,\mathbf{B}^T\mathbf{i}^{(l)} \tag{11}$$

Each of the $l$ elements of the column vector $\mathbf{Be}$ is the total e.m.f. in the direction of a loop. If these are the elements of the column vector $\mathbf{E}$, equation (11) may be written

$$\mathbf{Q}\mathbf{i}^{(l)} = \mathbf{E} \tag{12}$$

$\mathbf{Q}$ is an $l \times l$ matrix termed a loop impedance matrix given by the expression

$$\mathbf{Q} = \mathbf{B}\mathbf{Z}\mathbf{B}^T \tag{13}$$

It is to be noted that whereas the nodal admittance matrix $\mathbf{P}$ is unique, there may be several loop impedance matrices depending on the choice of independent loops.

## THE RELATION BETWEEN $\mathbf{A}_a$ AND B

The introduction of nodal voltages and loop currents has been made on a physical basis. The branch voltages are defined in terms of the nodal voltages by equation (4), here repeated

$$\mathbf{u} = \mathbf{A}^T\mathbf{v}$$

The branch voltages must satisfy equations (3′). Hence

$$\mathbf{B}\mathbf{A}^T\mathbf{v} = 0 \tag{14}$$

The nodal voltages can be arbitrarily specified by altering the sources and component values. Hence equation (14) can only be satisfied if

$$\mathbf{B}\mathbf{A}^T = 0 \tag{15}$$

This also implies

$$\mathbf{A}\mathbf{B}^T = 0 \tag{16}$$

More rigorously, equations (15) and (16) may be proved from the definition of the two matrices and the concepts of nodal voltages and loop currents then appear mathematically.

# INDEX

Printed in Dunstable, United Kingdom

91221453R00100